MOSSES OF THE NORTHERN FOREST

A PHOTOGRAPHIC GUIDE

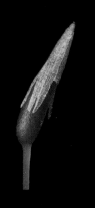

First published 2020 by Cornell University Press

Printed in China

Library of Congress Cataloging-in-Publication Data

Names: Jenkins, Jerry (Jerry C.), author.
Title: Mosses of the northern forest : a photographic guide / Jerry Jenkins.
Description: Ithaca [New York] : Comstock Publishing Associates, an imprint of Cornell University Press, 2020. | Series: A northern forest atlas guide | Includes bibliographical references and index.
Identifiers: LCCN 2019044577 | ISBN 9781501748615 (paperback)
Subjects: LCSH: Mosses—Northeastern States—Identification. | Mosses—Canada, Eastern—Identification. | Forest plants—Northeastern States—Identification. | Forest plants—Canada, Eastern—Identification.
Classification: LCC QK541 .J38 2020 | DDC 587/.90974—dc23
LC record available at https://lccn.loc.gov/2019044577

CONTENTS

QUICK GUIDES TO HABITATS

QUICK GUIDES TO ACROCARPS

QUICK GUIDES TO PLEUROCARPS

QUICK GUIDES TO SPHAGNUM

ACROCARPS

PLEUROCARPS

SPHAGNUM

Rhodobryum ontariense and *Thuidium recognitum*, Elwell Mill Cedar Swamp, Vermont

ABUNDANCE AND IDENTIFIABILITY

D In patches of several tens of centimeters or more; visible from a distance; overgrowing other mosses.

Dominant

C Easy to find within the proper habitat; in most patches of suitable habitat.

Common

U Hard to find within the proper habitat; small amounts, or absent from many patches of suitable habitat.

Uncommon

R Few sites in Northern Forest Region, or habitat very rare, or distribution poorly known.

Rare

 Identifiable with a hand lens, given familiarity; sometimes fruiting material is required.

Identifiable in field

 A common or dominant species of ecological significance that can be identified in the field.

Must know

 Microscope is needed, either to identify at all or to confirm a field identification.

INTRODUCTION

THE NORTHERN FOREST REGION (NFR) contains over 350 mosses. The number is imprecise because mosses are small, because relatively few people study them, and because the ones that do don't agree on how many they see.

This book contains photos, drawings, and brief descriptions of 305 mosses. These are the species that Sue Williams and I have seen and photographed in 25 years of mossing together. They include all the common species and many rare ones. In addition, we mention another 73 in the text or notes. These include rare species like *Bryum dichotomum* that we have never seen and problem species like *Atrichum crispulum* that we don't understand. With these exceptions, if we found it and believed in got a photo and a full entry. If we didn't or couldn't, it went in the notes.

Our goal, as in all the books of this series, is characterization and comparison. We want to show you what each species looks like, what other species it resembles, and what you need to see to identify it. Because mosses are small and hard, we can't promise that you will be able to identify everything you see. Many mosses need to be studied indoors, read about, argued about. Some, every mosser learns, are best admired and left alone.

The challenge in learning mosses is to know which are field identifiable and which aren't. In habitats we know well, we can identify 50% or more of the mosses with certainty in the field and make good guesses for another 30% or 40%. The big ones are mostly easy, the small ones often hard. Wise mossers start big.

To suggest good starting points, we use three symbols. ⓕ means that a species or genus is field identifiable; ᴹ, "must-know," means that a species is both field identifiable and common enough that it is a species that everyone should know; and ê means that a microscope is required for identification or confirmation.

Field identification of mosses is based on habitats and features. We lay this out in the "Quick Guides," which show the mosses in particular habitats or with particular features. There is, for example, a guide to the common mosses of limy boulders, one to the mosses with needle-tipped leaves, and one to the mosses with flattened, fernlike leaves.

Once you have a moss, the guides to habitats will show you the common species that it might be. The guides to features will eliminate a lot of these and leave you with a small list of probable matches. In identification, elimination is key: a naturalist's ability to recognize the odd and the rare often depends, as Holmes emphasized, on her ability to eliminate everything else.

After elimination comes confirmation. This is best done in the systematic sections, which have more detail than the guides. If a moss is identifiable in the field, the notes and diagrams will show you how. If it needs a microscope, we use a ê to warn you and then tell you some of the microscope characters you will need to use.* If it has

close relatives with which it may be confused, or is one of the species systematists fight about in bryo-bars, we warn you of that too.

The mosses, and hence our accounts of them, divide into three groups: acrocarps, pleurocarps, and *Sphagnum*. Acrocarps occur everywhere, grow erect, don't branch much, form groves and mounds, and have capsules at the tip of the stem. Pleurocarps like moisture and shade, arch and sprawl, branch like crazy, make mats and thatches, and have capsules on short side branches. Sphagnums dominate the northern bogs and fens, have big heads, clustered branches, and make lawns, hummocks, and cushions.

Acrocarp

Pleurocarp

Sphagnum

The mosses also divide by abundance, which we show with the icons at the bottom of the preceding page. Knowing abundance is extremely useful: you can, for example, see *Plagiomnium cuspidatum* in almost every woods. You will, on the other hand, only find *Plagiomnium medium* or *rostratum* by going to the right sort of place, and then only by looking in the right spots within that place.

Specifying abundances turns out to be hard. So much depends on where you have been, how hard you have looked, and how good a looker you are. We have done our best, but the northern forest is a big place and twenty-five years are not nearly enough. If you know things we don't, please write us. Or come and visit, and we will have a glass and trade moss lore.

This book is deliberately minimal: lots of pictures and diagrams, not much text, many details left out. It is designed to be your first moss book, not your last. We have on our desks, and use almost every day, the *Mosses of Eastern North America, Maine Mosses,* and the (wonderful) *Flore des bryophytes du Québec-Labrador;* we also consult the moss volumes of the online *Flora of North America* constantly. Our website, northernforestatlas.org, has additional moss pictures and moss articles. We have put the best of these pictures in *Mosses of the Northern Forest—A Digital Atlas,* downloadable from the website.

The taxonomy used is also minimal. This is partly a matter of our own prejudices: we like names, but not every name and not to excess. We treat new species as hypotheses and ask for proof. It is also a matter of practicality. We are working botanists and need species concepts that match what we see in the field. Some taxa, say *Mnium thomsonii* or *Sphagnum subfulvum,* though slippery at times, seem essential. We couldn't describe what we see without them. Others, for example the segregates of *Sphagnum recurvum,* don't work for us, and so we can't recommend them to you.

Our friends often disagree. It's a big world and much more fun that way.

And that's it. There's more about sources and technique on page 164. Get a good hand lens and someone to learn with or from. Spend some time with the visual glossary. Then start wherever you are. The path, which is deep and goes far beyond names, begins there.

* A diagrammatic guide to laboratory identification (*The Northeastern Mosses: A Graphical Guide*) is available for download on our website.

VISUAL GLOSSARY: MOSS STRUCTURE

capsule

seta

sporophyte

perichaetial
leaves

sex organs
within
leaves

gametophyte

ACROCARP

sporophyte

short side
branch

gametophyte

PLEUROCARP

sporophyte

capsules

perichaetial leaves

head

gametophyte

branches
in cluster

SPHAGNUM

PROTONEMA WITH GAMETOPHYTES AND CAPSULES

brown
rhizoids

shoot
(stem +
leaves)

ACROCARP SHOOT

main stem

side branches
with branch
leaves

stem leaves with
paraphyllia

PLEUROCARP SHOOT WITH PINNATE BRANCHING

head

spreading
branches

descending branches   branch with branch leaves   stem leaves   whorl of fan-shaped stem leaves   terminal bud

SPHAGNUM

MOSS SPORES germinate to form a velvety green protonema. The leafy gametophytes arise from the protonema. They bear eggs in archegonia, hidden
among the perichaetial leaves, and sperm in antheridia, which may be hidden
or exposed in an antheridial cup. The sperm fertilize the eggs, the eggs develop
into sporophytes. The sporophytes remain attached to the gametophytes.

THE SPHAGNUMS are an old, isolated group of mosses that specialize in stor-
ing water in large empty cells (*clear cells*) that open to the outside through
pores. They have erect stems that bear branches in clusters, a head of young
branches at the top of the stem, and stem leaves, which often differ greatly
from the branch leaves.

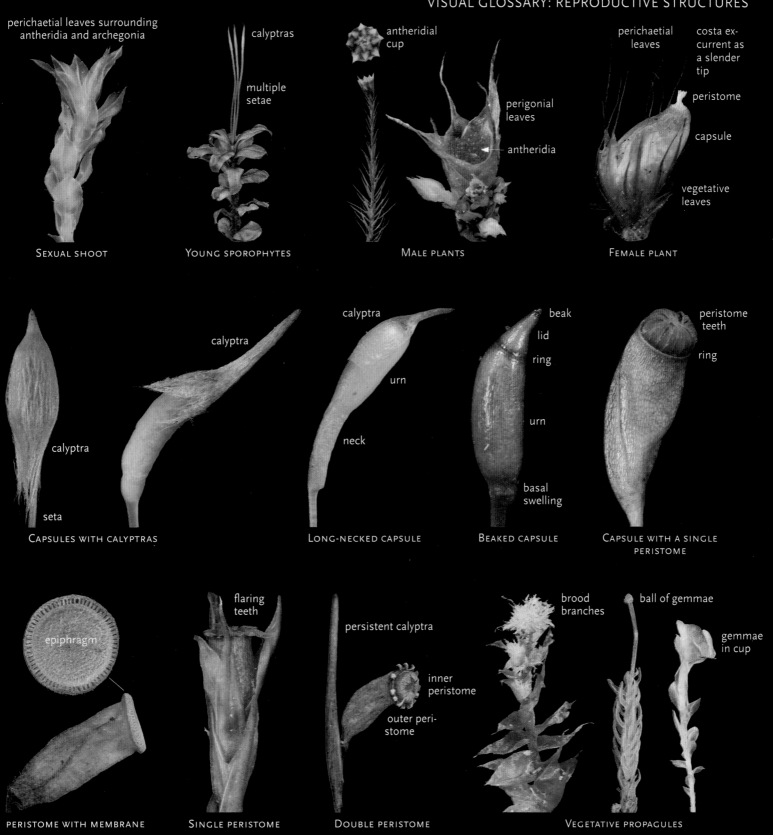

Sexual shoot — perichaetial leaves surrounding antheridia and archegonia

Young sporophytes — calyptras, multiple setae

Male plants — antheridial cup, perigonial leaves, antheridia

Female plant — perichaetial leaves, costa excurrent as a slender tip, peristome, capsule, vegetative leaves

Capsules with calyptras — calyptra, calyptra, calyptra, seta

Long-necked capsule — calyptra, urn, neck

Beaked capsule — beak, lid, ring, urn, basal swelling

Capsule with a single peristome — peristome teeth, ring

Peristome with membrane — epiphragm

Single peristome — flaring teeth

Double peristome — persistent calyptra, inner peristome, outer peristome

Vegetative propagules — brood branches, ball of gemmae, gemmae in cup

THE ANTHERIDIA of mosses may be on the same plant as the archegonia, either near the archegonia or on separate branches. If so, the plants are bisexual and the species monoicous. Or they may be on separate plants: if so, the plants are unisexual and the species dioicous. In the latter case the antheridia are often borne in cups at the tips of stems and the sperm dispersed by splashing

MOSS SPOROPHYTES usually consist of a seta and capsule. The capsule has an urn that holds the spores and, usually, a lid that falls off to release them. Under the lid is a peristome of one or two circles of teeth, which move with humidity changes and help eject the spores. *Sphagnum* capsules, which are blown apart by a charge of compressed air, have neither lids nor peristomes.

VISUAL GLOSSARY: LEAVES

OBLONG OVAL, WITH LAMELLAE ON COSTA

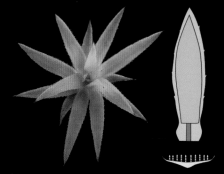

NEEDLELIKE, OPAQUE, COVERED BY LAMELLAE

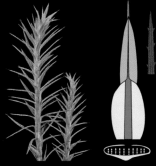

NEEDLELIKE, LAMELLAE COVERED BY EDGES

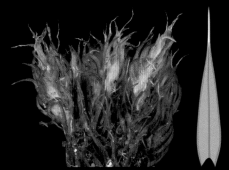

NARROW-LANCEOLATE, NEEDLE TIP

falcate-secund tips

COSTA EXCURRENT AS A SLENDER TIP

OBOVAL, BLUNT, BORDERED

NARROW, STRAPLIKE, BLUNT

OVAL, BORDERED

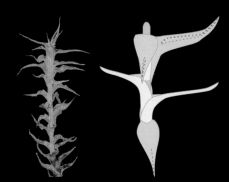

LANCEOLATE, SQUARROSE

TWO-ROWS, WITH A POUCH

ROUNDED-TRIANGULAR, PLEATED, WITH A DOUBLE COSTA

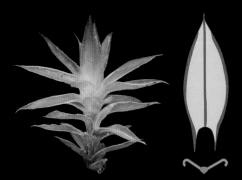

BORDERED, OVAL, DECURRENT

Moss LEAVES vary in shape, structural details, and the size and arrangement of the cells. The shape can be hard to see in the photos and is shown by icons that accompany each species. The icons also show costas and borders, which are structural elements, often thickened. Costas may be single, double, long, short, or absent. They may end below the leaf tip or be excurrent, running out as a slender point.

Borders may be thick, thin, toothed, or untoothed. Sometimes they can be seen with a lens; sometimes they need the microscope.

Leaf cells are necessary to identify many species and some genera. We list the critical characters in the species descriptions and, since they don't show in the photos, illustrate them with icons. The icons are largely self-explanatory. Some common ones are shown on page 5

HABITATS AND SUBSTRATES

Temperate forest

Boreal forest

Forest floor

Mixed swamp

On hummocks
in swamp

In hollows in
swamp

Open wetland

Boggy pool

Calcareous fen

Open bog

On bog hummocks

In bog hollows

Pool with *Sphagnum*

Wet bog mat

Soil bank

Seepy bank

Open soil

Wet sand

Wet ledge

Wet, dirty ledge

Seepy ledge

Calcareous seepage

Seepy bank

Limy rocks or soil

Alpine zone

Rocky barren

Forested ravine

Forested stream

Pond or lake

Rocks by water

Rocks by sea

Seepy shoreline ledges

Wood in water

Pond or stream shore

Pond in bog

Limy rocks by water

STRUCTURAL DETAILS

Stem and branch
leaves

Short double costa

Recurved edges

Incurved edges

Inflated alar cells

Paraphyllia on
leaf base

Oblong leaf, round cells

Ecostate leaf, long cells

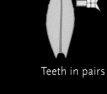
Teeth in pairs

Excurrent costa

Decurrent base

Papillose cells

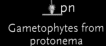
Gametophytes from
protonema

Perichaetial leaves,
lopsided capsule

Exserted capsule

Immersed capsule

Long-necked capsule

Long-beaked capsule

ABOUT MOSSES

Mosses are ancient spore-bearing plants, related to liverworts and hornworts. They were present over 300 million years ago; how much before and what they came from we don't know. Many of our modern genera are at least 50 million years old, making them among the oldest plants in the northern forest.

Like liverworts and hornworts, mosses alternate a leafy sexual generation (gametophyte) with a leafless spore-bearing one. The leafy gametophytes are the conspicuous, persistent parts of the plant. Structurally they are very simple: stems, leaves, containers for gametes, and assorted hairs and scales. No flowers, seeds, or fruits; very little plumbing. Vegetatively, however, they are quite diverse: large or small; erect, pendent, or creeping; branched or unbranched; and solitary or colonial.

Moss GAMETOPHYTES from the three main groups. From left, *Mnium hornum*, an acrocarp, with unbranched stems and oval leaves with teeth and borders. *Brachythecium populeum* and *Cirriphyllum piliferum*, pleurocarps with slender-pointed leaves and irregular (*Brachythecium*) or pinnate (*Cirriphyllum*) branching. *Sphagnum teres*, with the branches in clusters and a head of young branches.

ACROCARP LEAVES showing variations in shape and arrangement. From left, *Bartramia pomiformis*, linear-lanceolate leaves with long, curled tips; *Bryum pseudotriquetrum*, lanceolate leaves with borders, running down the stem at the base; and *Paludella squarrosa*, recurved lanceolate leaves in five rows.

Above, *Grimmia olneyi*, with white needle tips; *Polytrichum piliferum*, with opaque leaves covered by photosynthetic lamellae and white hair tips; and *Rhizomnium appalachianum*, broad-oval leaves bordered by clear cells. Like most acrocarps, all have costas.

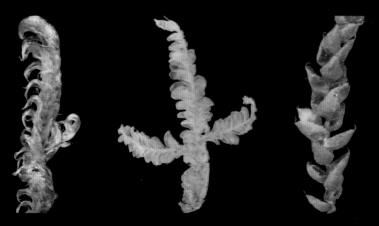

PLEUROCARP LEAVES, plus a common leafy liverwort for comparison. From left, *Hamatocaulis vernicosus*, linear-lanceolate leaves with the tips curled in circles; *Homalia trichomanoides*, blunt, oval leaves in two rows; and *Myurella sibirica*, deeply concave leaves with papillose (minutely bumpy) surfaces.

Above, *Campylium stellatum*, with channeled needle tips; *Plagiothecium laetum*, with flattened shoots and asymmetrical leaves without costas; and the liverwort *Bazzania trilobata*, with leaves in two rows which have small lobes at their tips. No moss has lobed leaves or leaves overlapping like this.

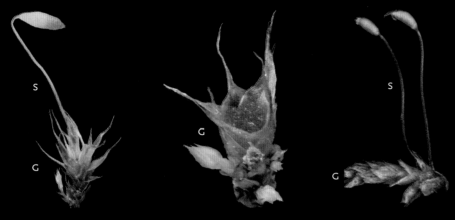

Moss gametophytes produce sperm and eggs from sex organs—antheridia and archegonia—which are borne at the tip of the main stem in acrocarps, and on short side branches in pleurocarps. Leaves surrounding the archegonia and antheridia are perichaetial and perigonial leaves. Some species have both sexes on one plant, others with antheridia on male plants and archegonia on female plants. The fertilized eggs produce sporophytes, consisting of a stalk (seta) and a capsule. The calyptra, which covers the developing capsule and regulates its development, is a remnant of the archegonia and so is a detached part of the sporophyte.

GAMETOPHYTES (G) AND SPOROPHYTES (S). Left, *Bryum lisae*, an acrocarp. The antheridia and archegonia are at the tip of the stem, hidden by the perichaetial leaves. Center, a male plant of *Philonotis marchica* with orange antheridia surrounded by a cup of perigonial leaves. Right, *Torrentaria riparioides*, a pleurocarp, with the setae arising from small side branches.

The capsules of most species open by a ring of specialized cells that separate the urn, which contains the spores, from the lid, which falls off. Within the lid are the peristome teeth. The teeth often move as the humidity changes, facilitating the release of spores.

ACROCARP CAPSULES vary greatly and are often important for identification. From left, *Polytrichum piliferum*, with a young capsule covered by the calyptra; *Tetraphis pellucida*, conical calyptra with lobed base; *Diphyscium foliosum*, lopsided capsule with conical lid and beak.

Above, *Schistidium apocarpum*, short capsule with flaring peristome, surrounded by enlarged perichaetial leaves; *Timmia megapolitana*, stubby capsule with white stomates and an erect calyptra; *Ulota hutchinsiae*, old capsules with reflexed peristomes, young capsule with hairy calyptra.

PLEUROCARP CAPSULES are much more uniform than acrocarp capsules and hence less useful for identification. Peristomes are largely double; size, stance and the presence or absence of beaks vary and are sometimes useful. But a lone pleurocarp capsule is, at best, difficult problem; often solvable with other

clues but not by itself. From left, capsules of *Anomodon rostratus*, *Anacamptodon splachnoides*, *Sematophyllum marylandicum* (showing the double peristome), *Brachythecium velutinum*, *Torrentaria riparioides*, *Plagiothecium cavifolium* (with a long-pointed calyptra), and *Callicladium haldanianum*.

MOSS MAP 1: LIVING TREES

ULOTA CRISPA

ULOTA COARCTATA

ORTHOTRICHUM SORDIDUM

THE GUIDES TO HABITATS include eleven moss maps, showing the commonest mosses in forty-one habitats. Where there is space we show the common liverworts as well, marked by asterisks to warn you that they are not included in the species accounts later in the book.

The maps focus on the common species, which are the ones you are likely to see, and which cover large areas. If you go, for example, into a woods with old trees and pick out the six or eight most conspicuous species on tree bark, the chance is very good that they will be on this page. If you then look hard and pick out another six species that are present only in small quantities or on a few trees, the chance is also good that they are uncommon species, treated later in the book but not shown on the maps.

LEUCODON ANDREWSIANUS

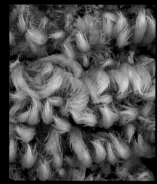

NECKERA PENNATA

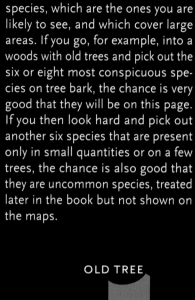

PYLAISIA SELWYNII

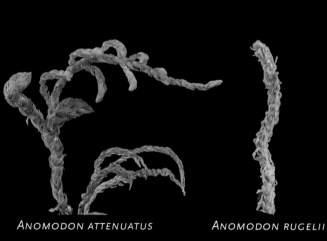

PORELLA PLATYPHYLLA*

ANOMODON ATTENUATUS

ANOMODON RUGELII

OLD TREE

Ulota

Orthotrichum

Platygyrium

Pylaisia

Neckera

Leucodon

Porella*

Hypnum

Anomodon

Brachythecium

Brotherella

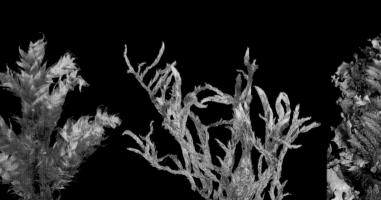

BROTHERELLA RECURVANS

BRACHYTHECIUM LAETUM

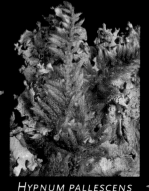

HYPNUM PALLESCENS

* A leafy liverwort, not in the species accounts.

YOUNG TREES, with small crowns and tight bark, have limited stem flow and storage. They are colonized by a relatively small group of pioneer species that make thin mats and small dense tufts. As the crown increases and the bark roughens, storage and flow increase, and the pioneers are supplemented by species that make fringes and shaggy mats on the trunk and thick stockings covering the base. In big old trees, the fringes and mats may extend high into the crown, and the tree base accumulates soil and becomes an extension of the forest floor.

Besides the species shown, trees may have other Anomodons Brachytheciums, Dicranums, and Orthotrichums; forest-floor genera like *Plagiomnium*, *Thuidium*, and *Hypnum*; and less common genera like *Trichostomum*, *Rauiella*, *Schwetschkeopsis*, *Platydictya*, and *Zygodon*.

ORTHOTRICHUM STELLATUM

*FRULLANIA EBORACENSIS**

ULOTA CRISPA

PLATYGYRIUM REPENS

YOUNG TREE

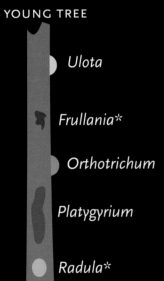

Ulota

*Frullania**

Orthotrichum

Platygyrium

*Radula**

Dicranum

Leskea

Callicladium

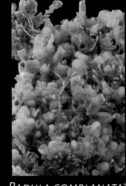

*RADULA COMPLANATA**

DICRANUM MONTANUM

CALLICLADIUM HALDANIANUM

LESKEA POLYCARPA GROUP

K GUIDES TO ABITATS

QUICK GUIDES TO ACROCARPS

QUICK GUIDES TO PLEUROCARPS

QUICK GUIDES TO SPHAGNUM

ACROCARPS

PLEUROCARPS

SPHAGNUM

ORTHOTRICHUM SORDIDUM

ULOTA CRISPA

PLATYGYRIUM REPENS

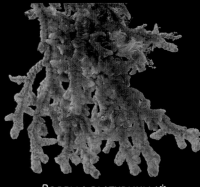

PORELLA PLATYPHYLLA*

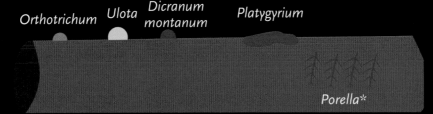

Orthotrichum   *Ulota*   *Dicranum montanum*   *Platygyrium*

*Porella**

RECENTLY FALLEN, BARK STILL ON, WITH TREE-TRUNK SPECIES

FALLEN LOGS gradually lose the species from the living tree, at least in part because they no longer have stem flow. Some tree species—*Dicranum montanum, Porella, Leskea*—may persist or flourish on logs.

Once the bark is gone and the log starts to soften, it is colonized, often rapidly, by a group of dead-wood specialists that form thin mats or small tufts. These may live on rocks as well but only rarely on the forest floor.

DICRANUM MONTANUM

PTILIDIUM PULCHERRIMUM*

DICRANUM FLAGELLARE

Dicranum flagellare   *Dicranum montanum*   *Ptilidium**

Hypnum pallescens

*Nowellia**

*Lophocolea**   *Brachythecium reflexum*

*Radula**

BARK IS GONE: TUFTS AND THIN MATS

HYPNUM PALLESCENS

NOWELLIA CURVIFOLIA*

BRACHYTHECIUM REFLEXUM

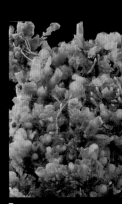

RADULA COMPLANATA

* A leafy liverwort, not in the species accounts

DICRANUM VIRIDE

HYPNUM IMPONENS

ENTODON CLADORRHIZANS

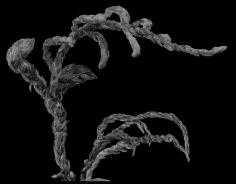

ANOMODON ATTENUATUS

CALLICLADIUM HALDANIANUM

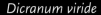

Dicranum viride

Callicladium *Hypnum imponens* *Entodon* *Anomodon*

Brotherella, Brachythecium, Plagiomnium, Thuidium

DECAYED LOG, SOFT AND WET, WITH THICK-MAT SPECIES;
FOREST-FLOOR SPECIES INVADING FROM BELOW

BROTHERELLA RECURVANS

TETRAPHIS PELLUCIDA

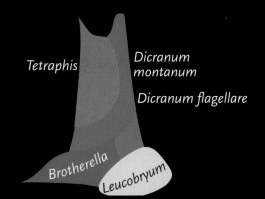

Tetraphis *Dicranum montanum*

Dicranum flagellare

Brotherella *Leucobryum*

DECAYED STUMP, DEAD-WOOD SPECIALISTS
ABOVE, FOREST-FLOOR SPECIES BELOW

DICRANUM FLAGELLARE

WHEN A LOG BECOMES DECAYED and
wet, it is colonized by large, thick-mat
species that overgrow and eliminate
the initial colonists and then try to do
the same to each other. As the log gets
softer and wetter and then starts to col-
lapse and become soil, these are joined,
or replaced, by common forest-floor
dominants: *Brachythecium, Thuidium,
Plagiothecium* and its relatives.

LEUCOBRYUM GLAUCUM

BRACHYTHECIUM CAMPESTRE

QUICK GUIDES TO HABITATS

QUICK GUIDES TO ACROCARPS

QUICK GUIDES TO PLEUROCARPS

QUICK GUIDES TO SPHAGNUM

ACROCARPS

PLEUROCARPS

SPHAGNUM

POLYTRICHUM JUNIPERINUM

POLYTRICHUM PILIFERUM

HEDWIGIA CILIATA

ULOTA HUTCHINSIAE

DICRANUM SCOPARIUM

SCHISTIDIUM APOCARPUM GROUP

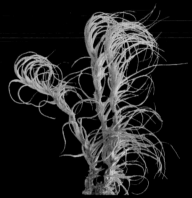

PARALEUCOBRYUM LONGIFOLIUM

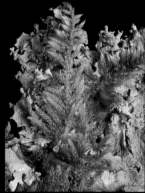

HYPNUM PALLESCENS

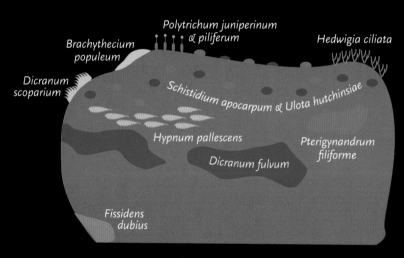

BOULDER IN DRY FOREST

DICRANUM FULVUM

FISSIDENS DUBIUS

BOULDERS receive most of their water from humidity and rain. When they are in the open, or in dry woods, they are among our driest habitats. Only a limited number of mosses can colonize them: the ones that do are fairly specialized: mostly acrocarps, often needle tipped, and often quite good at curling up to prevent water loss. They may occur on open soil and tree bases as well but, with exception of *Hypnum pallescens*, only rarely on logs or rocks.

PTERIGYNANDRUM, p. 143

12

QUICK GUIDES TO HABITATS

QUICK GUIDES TO ACROCARPS

QUICK GUIDES TO PLEUROCARPS

QUICK GUIDES TO SPHAGNUM

ACROCARPS

PLEUROCARPS

SPHAGNUM

DICRANUM SCOPARIUM

POLYTRICHUM PALLIDISETUM

POLYTRICHUM COMMUNE

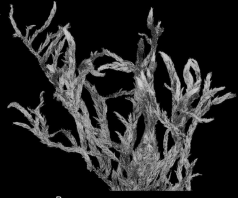

BRACHYTHECIUM LAETUM

DICRANUM MONTANUM

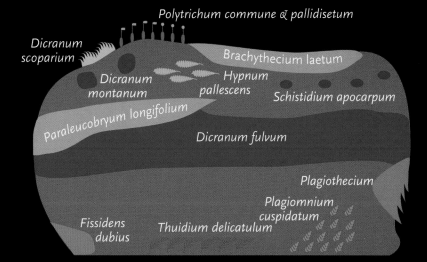

Polytrichum commune & pallidisetum

Dicranum scoparium

Dicranum montanum

Brachythecium laetum

Hypnum pallescens

Schistidium apocarpum

Paraleucobryum longifolium

Dicranum fulvum

Plagiothecium

Plagiomnium cuspidatum

Fissidens dubius

Thuidium delicatulum

BOULDER IN MESIC FOREST

DICRANUM FULVUM

FISSIDENS DUBIUS

THUIDIUM DELICATULUM

PLAGIOMNIUM CUSPIDATUM

PLAGIOTHECIUM

BOULDERS ARE COUPLED to the surrounding forest by through-fall, litter-fall, and ground-layer humidity. The wetter the forest, the wetter and dirtier the boulder. In mesic forests with closed canopies, the xeric mosses like *Ulota*, *Hedwigia*, and *Polytrichum piliferum* drop out. Soil accumulates on top of the boulders and thick-mat species like the Brachytheciums and tall grove-forming species like the forest Polytrichums and Dicranums dominate. *Dicranum fulvum* expands to make thick, continuous mats; the Plagiotheciums, which specialize in flow interception, become common on the sides. Humidity-requiring species like *Thuidium* and *Plagiomnium* from the forest floor colonize the lower sides.

DICRANUM SCOPARIUM

DICRANUM POLYSETUM

PTILIUM CRISTA-CASTRENSIS

POLYTRICHUM COMMUNE

HYLOCOMIUM SPLENDENS

*BAZZANIA TRILOBATA**

HYPNUM IMPONENS

PLEUROZIUM SCHREBERI

Dicranum
scoparium

Dicranum
polysetum

Hylocomium
splendens

Ptilium
crista-castrensis

Pleurozium
schreberi

Ptilidium
ciliare*

Bazzania trilobata*

Hypnum
imponens

Dicranum
polysetum

Polytrichum
commune

Hypnum
imponens

Leucobryum
glaucum

*PTILIDIUM CILIARE**

MOIST BOREAL FORESTS often have a continuous moss layer within about one meter of the forest floor. Acrocarps and pleurocarps are thoroughly mixed and often compete on equal terms. Acrocarps make mounds or groves; pleurocarps make thatches and mats. Most species grow loosely to allow conifer litter to percolate through them. The litter accumulates within the moss layer and forms an organic soil. The presence of this soil blurs the differences between substrates. Rocks and logs become extensions of the forest floor. Given litter and humidity, the main species, which we call *coveralls*, will grow on almost anything.

LEUCOBRYUM GLAUCUM

* A leafy liverwort, not in the species accounts.

QUICK GUIDES TO HABITATS

QUICK GUIDES TO ACROCARPS

QUICK GUIDES TO PLEUROCARPS

QUICK GUIDES TO SPHAGNUM

ACROCARPS

PLEUROCARPS

SPHAGNUM

TORTELLA TORTUOSA

RHODOBRYUM ONTARIENSE

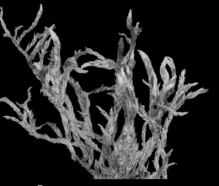

BRACHYTHECIUM LAETUM

HOMOMALLIUM ADNATUM

ANOMODON VITICULOSUS

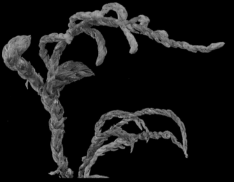

ANOMODON ATTENUATUS

FISSIDENS SUBBASILARIS

AMBLYSTEGIUM VARIUM

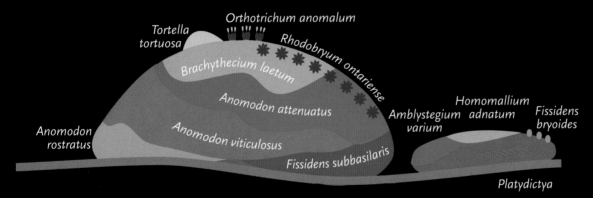

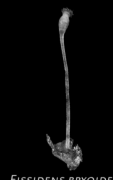

FISSIDENS BRYOIDES

ANOMODON ROSTRATUS

LIMY BOULDERS in moist forests are dominated by a specialized group of pleurocarps that make thick mats, often covering the entire rock and excluding other species. Three Anomodons and one or more Brachytheciums are the common dominants. *Rhodobryum* is a common associate. Other acrocarps, particularly species of *Fissidens* and *Tortella*, are also common.

Smaller rocks are much dryer, and typically have small acrocarps—*Fissidens, Schistidium, Tortella*—and small, stringy, thin-mat pleurocarps. *Amblystegium* and *Homomallium* are the commonest; *Platydictya*, our smallest pleurocarp, is a frequent associate.

PLATYDICTYA CONFERVOIDES

POLYTRICHUM JUNIPERINUM

POLYTRICHUM PILIFERUM

GRIMMIA PILIFERA

HEDWIGIA CILIATA

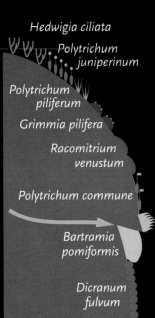

Hedwigia ciliata

Polytrichum juniperinum

Polytrichum piliferum

Grimmia pilifera

Racomitrium venustum

Polytrichum commune

Dicranum scoparium

Bartramia pomiformis

Dicranum fulvum

Polytrichastrum alpinum

Bryum capillare

Herzogiella striatella

Brachythecium laetum

DRY LEDGE WITH SEEPAGE

RACOMITRIUM VENUSTUM

DICRANUM SCOPARIUM

POLYTRICHUM COMMUNE

BRYUM CAPILLARE

BARTRAMIA POMIFORMIS

DICRANUM FULVUM

ABOVE, a cliff that is sunny at the top and shaded at the base. The top receives surface flow, the lower parts surface flow and seepage. Most of the cliff is dominated by acrocarps, with xeric grove and dense-cushion species above and larger mat and loose-cushion species below. Near the forest floor, where the humidity is highest, thatch-forming species can establish.

POLYTRICHASTRUM ALPINUM

BRACHYTHECIUM LAETUM

QUICK GUIDES TO HABITATS

QUICK GUIDES TO ACROCARPS

QUICK GUIDES TO PLEUROCARPS

QUICK GUIDES TO SPHAGNUM

ACROCARPS

PLEUROCARPS

SPHAGNUM

BRYUM PSEUDOTRIQUETRUM

ANDREAEA

*SCAPANIA NEMOREA**

SCHISTIDIUM APOCARPUM GROUP

BRACHYTHECIUM RIVULARE

SPHAGNUM GIRGENSOHNII

PLAGIOTHECIUM

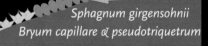

Andreaea
Dicranum fulvum
Schistidium apocarpum

Sphagnum girgensohnii

Scpania nemorea*

Philonotis fontana

Plagiothecium, Pseudotaxiphyllum

Rhabdoweisia crispata

Metzgeria*

Sphagnum girgensohnii
Bryum capillare & pseudotriquetrum

Brachythecium rivulare
& Bryhnia novae-angliae

Fissidens dubius, Pseudotaxiphyllum

WET LEDGE WITH SURFACE FLOW

BRYHNIA NOVAE-ANGLIAE

RHABDOWEISIA CRISPATA

*METZGERIA**

WET CLIFFS are good moss habitats in the summer and bad ones in the winter. The species sort by how much water they need and how much ice damage they can tolerate. Small delicate species live under overhangs. Large somewhat tolerant ones live on slopes where they get frozen but not stripped away. Small tough mat-formers and opportunists that regenerate quickly live on the vertical faces where the ice damage is worst.

FISSIDENS DUBIUS

PSEUDOTAXIPHYLLUM DISTICHACEUM

* A leafy liverwort, not in the species accounts.

17

Rhytidium rugosum

Thuidium recognitum

Abietinella abietina

Hypnum cupressiforme

Schistidium apocarpum group

Tortella tortuosa

Rhodobryum ontariense

Fissidens subbasilaris

Anomodon attenuatus

Anomodon viticulosus

Encalypta procera

Mnium marginatum

Anomodon rostratus

Tortella tortuosa
Rhytidium rugosum
Thuidium recognitum

Schistidium apocarpum

Rhodobryum ontariense

Anomodon attenuatus

Hypnum cupressiforme

Open limy knoll

Fissidens subbasilaris

Mnium marginatum
& thomsonii

Anomodon viticulosus

Encalypta procera Barbula

Anomodon rostratus

QUICK GUIDES TO HABITATS

QUICK GUIDES TO ACROCARPS

QUICK GUIDES TO PLEUROCARPS

QUICK GUIDES TO SPHAGNUM

ACROCARPS

PLEUROCARPS

SPHAGNUM

BARBULA CONVOLUTA

TIMMIA MEGAPOLITANA

GYMNOSTOMUM AERUGINOSUM

BRYOERYTHROPHYLLUM RECURVIROSTRUM

MYURELLA SIBIRICA

POHLIA CRUDA

AMPHIDIUM

BARTRAMIA POMIFORMIS

PLAGIOPUS OEDERIANA

PLAGIOMNIUM ROSTRATUM

LIMY KNOLLS AND LEDGES have a special-ized flora not found elsewhere. Dry open knolls have drought-tolerant mound-builders and thatch-makers like *Tortella, Abietinella,* and *Rhytidium;* moister ledges have deep mats of *Anomodon*; dripping faces have *Plagiopus, Bartramia,* and *Timmia*. Seepage cracks and moist crevices have a group of small, often northern, species that can't compete out in the open: *Myurella, Pohlia, Bryoerythrophyllum, Gymnostomum, Didy-modon,* and *Barbula*.

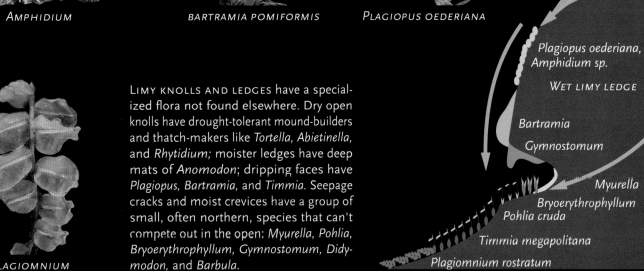

Plagiopus oederiana, Amphidium sp.

WET LIMY LEDGE

Bartramia

Gymnostomum

Myurella

Bryoerythrophyllum
Pohlia cruda

Timmia megapolitana

Plagiomnium rostratum

Plagiomnium ciliare

Mnium hornum

Bryhnia novae-angliae

Brachythecium plumosum

Platylomella lescurii

Fissidens bryoides

Atrichum undulatum group

Hygrohypnum eugyrium

Climacium

Fontinalis antipyretica

Hygroamblystegium tenax

Torrentaria riparioides

Thuidium delicatulum

Atrichum undulatum

- - - - - - - - - - high water - - - - - -

Mnium hornum, Plagiomnium ciliare — *Climacium*

ROCKS IN CHANNEL

Sematophyllum marylandicum

Fissidens bryoides

Bryhnia novae-angliae

Brachythecium plumosum

Hygroamblystegium

MINERAL-SOIL BANKS

Platylomella lescurii

*Scapania**

Hygrohypnum

Fontinalis

Torrentaria riparioides

QUICK GUIDES TO HABITATS

QUICK GUIDES TO A-RCCARPS

QUICK GUIDES TO PLEUROCARPS

QUICK GUIDES TO SPHAGNUM

ACROCARPS

PLEUROCARPS

SPHAGNUM

THAMNOBRYUM ALLEGHANIENSE

AMPHIDIUM

PLAGIOTHECIUM

FISSIDENS DUBIUS

RHIZOMNIUM PUNCTATUM

SCHISTIDIUM RIVULARE

RACOMITRIUM

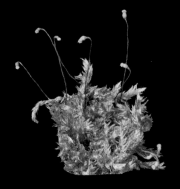

SEMATOPHYLLUM MARYLANDICUM

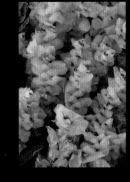

*SCAPANIA NEMOREA**

*CONOCEPHALUM CONICUM**

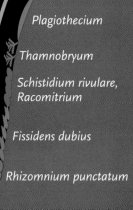

Amphidium

Plagiothecium

Thamnobryum

Schistidium rivulare, Racomitrium

LEDGE IN
SPLASH ZONE

Fissidens dubius

– – – – high water – – – –

Rhizomnium punctatum

Conocephalum conicum

Scapania*

Hygrohypnum, Torrrentaria

ROCKY, SHADED STREAMS have three overlapping groups of bryophytes: forest-floor species, wet-ledge species, and stream-channel species. The forest-floor species, typified by *Mnium hornum* and *Climacium dendroides*, grow on the banks where sand is deposited and groundwater surfaces. The wet-ledge species, typified by *Schistidium* and *Thamnobryum*, benefit from spray and seepage; many are minerotrophs. The stream channel specialists, typified by *Hygrohypnum* and *Torrentaria*, live in the active channel, mostly between the average high-water and low-water levels. All three groups have species that can tolerate submersion; the channel specialists may require it.

* A leafy liverwort, not in the species accounts.

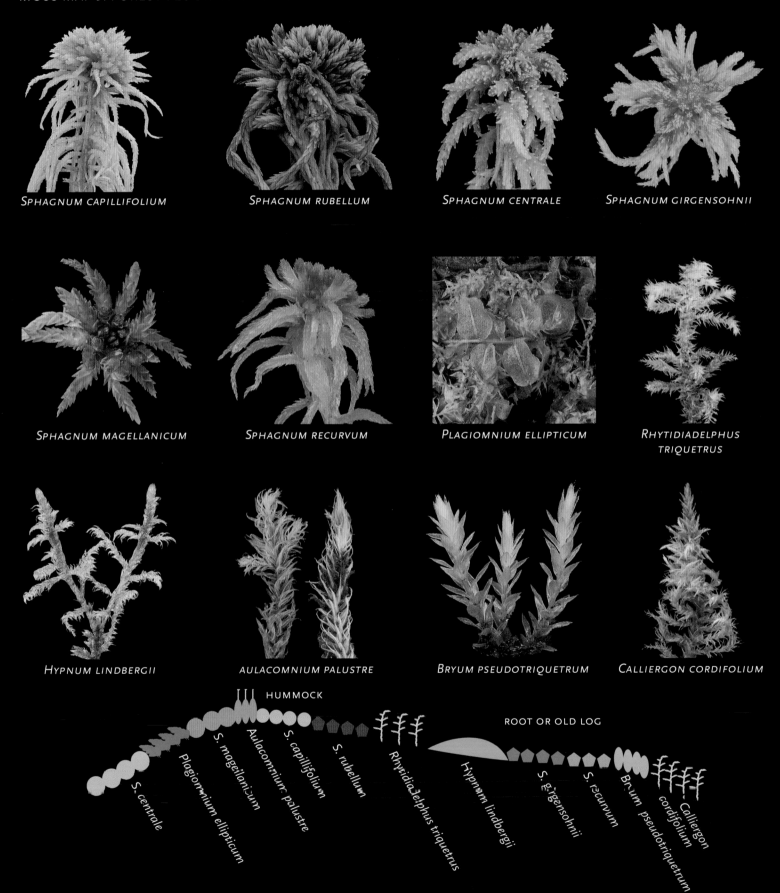

SPHAGNUM CAPILLIFOLIUM

SPHAGNUM RUBELLUM

SPHAGNUM CENTRALE

SPHAGNUM GIRGENSOHNII

SPHAGNUM MAGELLANICUM

SPHAGNUM RECURVUM

PLAGIOMNIUM ELLIPTICUM

RHYTIDIADELPHUS TRIQUETRUS

HYPNUM LINDBERGII

AULACOMNIUM PALUSTRE

BRYUM PSEUDOTRIQUETRUM

CALLIERGON CORDIFOLIUM

HUMMOCK

ROOT OR OLD LOG

S. centrale

Plagiomnium ellipticum

S. magellanicum

Aulacomnium palustre

S. capillifolium

S. rubellum

Rhytidiadelphus triquetrus

Hypnum lindbergii

S. girgensohnii

S. recurvum

Bryum pseudotriquetrum

Calliergon cordifolium

* Swamps will also have many of the common genera of trees, logs, and the boreal forest floor, particularly *Brotherella, Climacium, Dicranum, Hylocomium, Hypnum, Leucobryum, Pleurozium, Ptilium, Orthotrichum, Thuidium,* and *Ulota.* For these see Maps 1, 2 and 4

QUICK GUIDES TO HABITATS

QUICK GUIDES TO ACROCARPS

QUICK GUIDES TO PLEUROCARPS

QUICK GUIDES TO SPHAGNUM

ACROCARPS

PLEUROCARPS

SPHAGNUM

MOSS MAP 9: FOREST FLOOR IN A MEDIUM-FERTILITY SWAMP

RHIZOMNIUM APPALACHIANUM

PSEUDOBRYUM CINCLIDIOIDES

RHYTIDIADELPHUS SQUARROSUS

DICRANUM VIRIDE

LEPTODICTYUM RIPARIUM

BRYHNIA NOVAE-ANGLIAE

SPHAGNUM SQUARROSUM

ONCOPHORUS WAHLENBERGII

HELODIUM PALUDOSUM

CAMPYLIUM STELLATUM

SPHAGNUM PALUSTRE

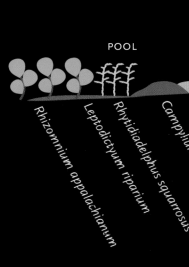

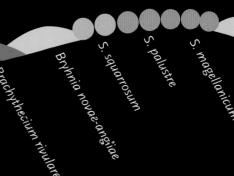

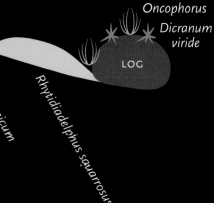

HUMMOCK

Oncophorus

Dicranum viride

POOL

LOG

Rhizomnium appalachianum

Leptodictyum riparium

Rhytidiadelphus squarrosus

Campylium stellatum

Helodium paludosum

Pseudobryum cinclidioides

Brachythecium rivulare

Bryhnia novae-angiae

S. squarrosum

S. palustre

S. magellanicum

Rhytidiadelphus squarrosus

Sphagnum magellanicum　　*Sphagnum russowii*　　*Sphagnum capillifolium*　　*Sphagnum rubellum*

Sphagnum fimbriatum　　*Sphagnum cuspidatum*　　*Sphagnum majus*　　*Sphagnum torreyanum*＊

Sphagnum subsecundum　　*Sphagnum papillosum*　　*Sphagnum tenellum*＊

russowii

subsecundum

magellanicum
papillosum
fimbriatum

majus, cuspidatum,
tenellum＊, torreyanum＊

capillifolium
rubellum

recurvum,
pulchrum＊

MARSHY
SHORE

BOG POND

FLOATING MAT

POOL

HUMMOCK

LAWN

Bogs are open or sparsely wooded rain-fed peatlands; they are acid, nutrient poor, and invariably dominated by a small group of Sphagnums. We show 13 species of open bogs here; not all of these will occur in any one bog. The species divide by ecology. The hummocks are dominated by species of Section *Acutifolia*: *fuscum* at the top, *capillifolium* below that, *rubellum* below that. Species of Section *Cuspidata* dominate the lawns and pools: the *recurvum* group in the lawns and on the sides of the hummocks, *pulchrum* in low wet lawns, and *majus* and *cuspidatum* in the pools. Section *Sphagnum*, which tends to be more minerotrophic, is especially common on wet mats: *magellanicum* occurs on hummocks, and *papillosum* is a common dominant in floating mats. *Fimbriatum*, *russowii*, and *subsecundum*, all weak minerotrophs, occur on shores, in alder tickets, and in places with groundwater seepage.

＊ The three species with asterisks are mostly found near the Atlantic coast. *Flavicomans*, in Section *Acutifolia*, grows with and below *fuscum* in hummocks. *Tenellum* and *torreyanum*, in Section *Cuspidata*, are semiaquatics found in pools.

DICRANUM SCOPARIUM

POLYTRICHUM STRICTUM

AULACOMNIUM

SPHAGNUM CAPILLIFOLIUM

SPHAGNUM FUSCUM

SPHAGNUM FLAVICOMANS*

SPHAGNUM RUBELLUM

SPHAGNUM GIRGENSOHNII

SPHAGNUM CENTRALE

SPHAGNUM MAGELLANICUM

SPHAGNUM PULCHRUM

SPHAGNUM RECURVUM GROUP

Aulacomnium palustre   *fuscum, flavicomans**
Dicranum scoparium   *capillifolium, rubellum*
Polytrichum strictum   *recurvum*

capillifolium, centrale
magellanicum
girgensohnii
recurvum

recurvum,
pulchrum*

LAWN   HUMMOCK

OPEN BOG   BOG THICKET   SPRUCE-TAMARACK SWAMP

VERY FEW MOSSES OTHER THAN *SPHAGNUM* occur in open bogs. *Poly-trichum strictum* is common at the tops of dry hummocks; *Dicranum scoparium* and *Aulacomnium palustre* are occasional and *Dicranum undulatum* rare.

BLACK SPRUCE-TAMARACK SWAMPS and bog thickets are partially forested bogs with slightly more nutrients. They often have a contin-uous mat of *Sphagnum*. *Recurvum* and *girgensohnii* are the common dominants. Green forms of *capillifolium* are common on hummocks. Green forms of *magellanicum* occur and, if the nutrient level is high enough, *centrale*, a minerotrophic species of Section *Sphagnum*, may be common. Other boreal forest species—*Hylocomium splendens*, *Hypnum imponens*, the liverwort *Bazzania trilobata*—occur on logs and stumps. *Calliergon cordifolium* and *Rhizomnium appalachianum* may occur in pools.

QUICK GUIDES TO HABITATS

QUICK GUIDES TO ACROCARPS

QUICK GUIDES TO PLEUROCARPS

QUICK GUIDES TO SPHAGNUM

ACROCARPS

PLEUROCARPS

SPHAGNUM

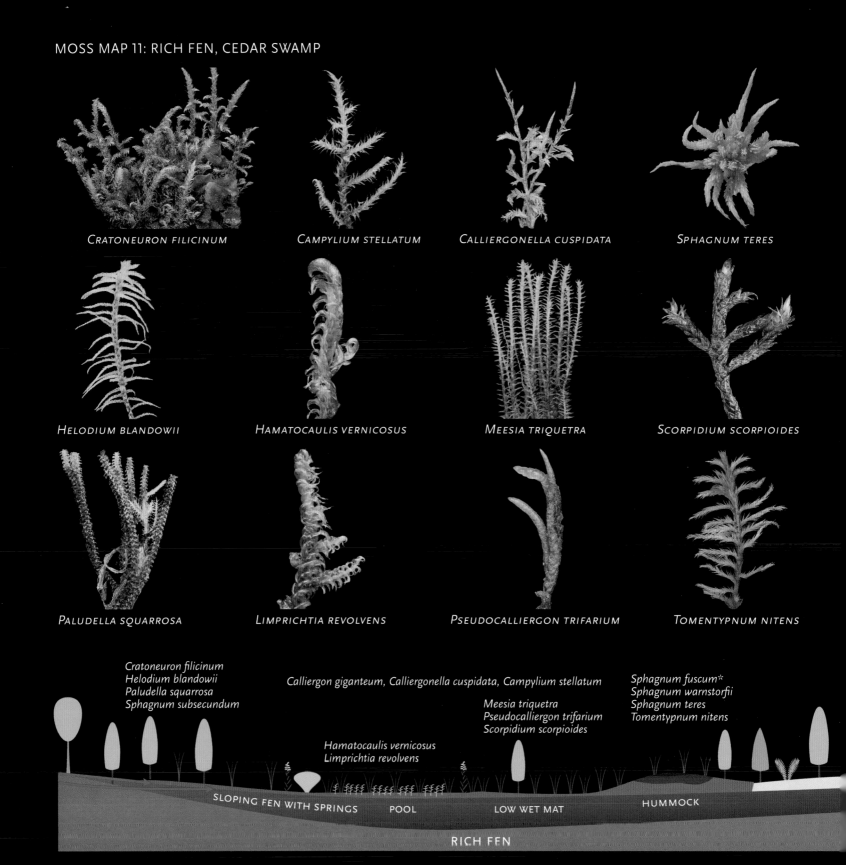

CRATONEURON FILICINUM

CAMPYLIUM STELLATUM

CALLIERGONELLA CUSPIDATA

SPHAGNUM TERES

HELODIUM BLANDOWII

HAMATOCAULIS VERNICOSUS

MEESIA TRIQUETRA

SCORPIDIUM SCORPIOIDES

PALUDELLA SQUARROSA

LIMPRICHTIA REVOLVENS

PSEUDOCALLIERGON TRIFARIUM

TOMENTYPNUM NITENS

Cratoneuron filicinum
Helodium blandowii
Paludella squarrosa
Sphagnum subsecundum

Calliergon giganteum, Calliergonella cuspidata, Campylium stellatum

Sphagnum fuscum*
Sphagnum warnstorfii
Sphagnum teres
Tomentypnum nitens

Meesia triquetra
Pseudocalliergon trifarium
Scorpidium scorpioides

Hamatocaulis vernicosus
Limprichtia revolvens

SLOPING FEN WITH SPRINGS   POOL   LOW WET MAT   HUMMOCK

RICH FEN

RICH FENS ARE OPEN PEATLANDS, typically dominated by sedges and mosses, which receive minerals from limy groundwater. The commonest mosses are *Calliergon*, *Calliergonella*, *Campylium*, *Hamatocaulis*, and *Limprichtia*. Where limy water surfaces, *Cratoneuron* and *Helodium* are common dominants, and minerotrophic Sphagnums like *subsecundum* may occur. Where there are hummocks the common dominants are *Sphagnum fuscum* (deep brown), *Sphagnum warnstorfii* (glossy purple-red), and *Tomentypnum nitens* (golden).

Rich fens are widespread in the arctic and subarctic, and some far-northern species come south to the NFR. We show four of our favorites here: *Meesia triquetra*, *Paludella squarrosa*, *Pseudocalliergon trifarium*, and *Scorpidium scorpioides*.

QUICK GUIDES TO HABITATS

QUICK GUIDES TO ACROCARPS

QUICK GUIDES TO PLEUROCARPS

QUICK GUIDES TO SPHAGNUM

ACROCARPS

PLEUROCARPS

SPHAGNUM

Sphagnum centrale

Sphagnum girgensohnii

Sphagnum magellanicum

Sphagnum recurvum group

Sphagnum subfulvum

Sphagnum warnstorfii

Sphagnum wulfianum

Campylium chrysophyllum

Calliergon giganteum

Ctenidium molluscum

*Trichocolea tomentella**

Hylocomium umbratum

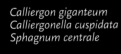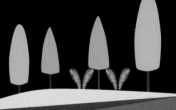

Sphagnum girgensohnii, magellanicum, recurvum

Calliergon giganteum
Calliergonella cuspidata
Sphagnum centrale

Sphagnum warnstorfii
Sphagnum subfulvum*

Sphagnum wulfianum
Trichocolea tomentella*
Campylium chrysophyllum
Ctenidium molluscum

Hylocomium
umbratum

POOL

HUMMOCK

TREE BASE OR STUMP   LOG

LIMY CONIFER SWAMP

WHITE CEDAR SWAMPS are typically more fertile than spruce-fir-tamarack or hemlock swamps. The dominant ground-layer mosses are typically a mixture of Sphagnums (*centrale, girgensohnii, magellanicum, recurvum* group, *warnstorfii*) with *Calliergon giganteum* and *Calliergonella cuspidata*. If they are limy enough, two uncommon species occur: *S. wulfianum*, which looks like green lollipops; and *S. subfulvum*, which looks like very little. Two frizzy bryophytes, the moss *Ctenidium* and the liverwort *Trichocolea*, are often common on hummocks. Many common forest bryos live higher up on logs and tree trunks. *Hylocomium umbratum*, an uncommon one, occurs regularly on logs in cedar swamps.

* *Trichocolea* is a leafy liverwort, not in the species accounts. For pictures of Sphagnum *fuscum* and *subsecundum*, see pages 24-25.

The NFR has about 70 genera of acrocarps. All have setae from the tips of the main stems. Most have sparsely branched stems and costate leaves. Within these boundaries, they are quite diverse: large or small, annual or perennial, flat or rounded, sleek or shaggy. Their capsules are similarly diverse: by using leaves and capsules it is possible to identify many genera with a hand lens.

To identify acrocarps, start with the guides to dominants, distinctive shoots, and distinctive leaves on p. 29-32. If you don't find your plant there, look in the ordinary leaves on p. 33-36 and the capsule guide on p. 37. When you find a genus that matches, check its features and ecology in the Systematic Section. Sometimes this will give you a firm ID. Other times it will give send you to the microscope scope and other books. Mosses are hard, and guesses are good.

TYPICAL ACROCARPS. From left, two years' growth of *Dicranum bonjeanii*; terminal setae and capsules of *Ulota hutchinsiae;* male plant with antheridial cup and young capsule covered by a silky calyptra of *Polytrichum commune*.

DISTINCTIVE SHOOTS AND CAPSULES. These will be found in the guides on pages 30 and 37. From left, *Meesia triquetra*, with squarrose (bent outward) leaves in three rows; *Anomobryum julaceum*, with pale-green leaves and cylindrical, wormlike shoots; and *Pogonatum pensilvanicum*, with the tiny gametophytes and sporophytes arising from a persistent protonema. Above, *Physcomitrium pyriforme*, a small, almost stemless annual; *Plagiomnium ciliare* with oblong oval leaves in two rows; and *Conostomum tetragonum*, a tiny blue-green moss forming cushions in the alpine zone, with the leaves in five rows.

DISTINCTIVE LEAVES. These will be found in the guides on pages 31 and 32. From left, *Pogonatum dentatum*, with opaque, needlelike leaves covered by photosynthetic ridges; *Rhizomnium appalachianum*, with blunt, broadly rounded leaves bordered by clear cells; and *Grimmia olneyi*, with lanceolate leaves with white needle tips. Above, *Tortella fragilis* with slender straight leaves with deciduous leaf tips; *Barbula unguiculata*, with slender, straplike leaves that are rounded at the tip and have a tiny point; and *Dicranella schreberiana*, with lanceolate leaves that arch outwards from erect bases.

ANDREAEA, p. 60   ATRICHUM, p. 62   BRYUM, p. 65   DICRANUM, p. 71

FUNARIA, p. 82   HEDWIGIA, p. 85   LEUCOBRYUM, p. 86   MNIUM, p. 87

PARALEUCOBRYUM, p. 91   PHILONOTIS, p. 92   PLAGIOMNIUM, p. 92   POGONATUM, p. 95

POLYTRICHUM, p. 98   RHIZOMNIUM, p. 103   TETRAPHIS, p. 107   TORTELLA, p. 107

DOMINANT ACROCARPS. These are the commonest acrocarps in the NFR. You will see them whenever you are in the proper habitat. *Atrichum, Polytrichum,* and *Pogonatum* have photosynthetic ridges (lamellae) on the costa. *Funaria,* with budlike, almost stemless shoots, is our commonest short-lived pioneer. *Dicranum, Paraleucobryum* and *Tortella* have elongate leaves that taper to long tips. *Tortella* and *Philonotis* are minutely papillose and become opaque when dry. *Bryum, Mnium, Plagiomnium,* and *Rhizomnium* have bordered leaves; *Plagiomnium* also has leaves in two rows. *Leucobryum* is white green, *Andreaea* almost black, and *Hedwigia* dark olive wet and light gray-green dry. All are benchmark species that you need to know. Learn them well, and you will be glad you did.

QUICK GUIDES TO HABITATS

QUICK GUIDES TO ACROCARPS

QUICK GUIDES TO PLEUROCARPS

QUICK GUIDES TO SPHAGNUM

ACROCARPS

PLEUROCARPS

SPHAGNUM

LEAVES IN 2 ROWS

CYRTOMNIUM, p. 69   *DISTICHIUM*, p. 77   *PLAGIOMNIUM*, p. 92   *FISSIDENS*, p. 81   *SCHISTOSTEGA PENNATA*, p. 105

LEAVES IN 3 ROWS

CATOSCOPIUM, p. 68   *MEESIA*, p. 86   *PLAGIOPUS*, p. 94

LEAVES IN 5 ROWS

CONOSTOMUM, p. 68   *PALUDELLA*, p. 91

SHORT STEMS OR STEMLESS

APHANORRHEGMA, p. 61   *EPHEMERUM*, p. 79   *BRYUM*, p. 65   *BUXBAUMIA*, p. 67   *FISSIDENS*, p. 81   *DIPHYSCIUM FOLIOSUM*, p. 76

SHORT STEMS OR STEMLESS

FUNARIA, p. 82   *PHYSCOMITRIUM*, p. 92   *POGONATUM*, p. 95   *SELIGERIA*, p. 105   *WEISSIA*, p. 111

DISTINCTIVE ACROCARPS. The next three pages show the acrocarps with unusual shoots and leaves. Check them first. If your plant isn't distinctive, check "Ordinary Acrocarps" on pages 33 to 36 for the remaining species. Note that the distinctive genera are not repeated in the ordinary sections: for example, *Plagiomnium*, with distinctive oval leaves in two rows, is not in the ordinary ovals on page 35.

DISTINCTIVE SHOOTS: A shoot is a stem with leaves. The mosses with distinctive shoots fall into three groups: those with the leaves in rows; those with the stems short and the leaves nearly basal; and those with the leaves broad and tightly overlapping, making cylindrical ("julaceous") shoots. The distinctions are straightforward, though sometimes hard to see in the field.

ACROCARPS WITH DISTINCTIVE SHOOTS

Anomobryum julaceum, p. 61

Bryum argenteum, p. 65

Plagiobryum zeiri, p. 92

ACROCARPS WITH DISTINCTIVE LEAVES

NEEDLE-TIPPED LEAVES

Encalypta rhaptocarpa, p. 79

Grimmia, p. 83

Hedwigia, p. 85

Pogonatum, p. 95

NEEDLE-TIPPED LEAVES

Polytrichastrum and *Polytrichum*, p. 98

Racomitrium, p. 100

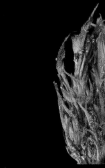

Schistidium, p. 104

Syntrichia, p. 106

PHOTOSYNTHETIC LAMELLAE

Atrichum, p. 62

Pogonatum, p. 95

Polytrichastrum, p. 98

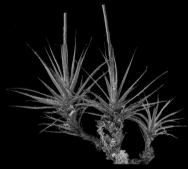

Polytrichum, p. 98

DISTINCTIVE LEAVES I. HERE, needle tips are stiff points that are demarcated from the leaf tissue below in color or texture. Photosynthetic lamellae are ridges of green tissue that arise from the costa and run the length of the leaf. In *Atrichum* they form a narrow band that can be seen with a lens. In *Polytrichum*, *Polytrichastrum*, and *Pogonatum*, they cover the entire width of the leaf and are hard to see in the field. Look for the thick opaque leaves instead. As a rule of thumb, moss leaves that look like spruce needles probably have lamellae.

QUICK GUIDES TO HABITATS

QUICK GUIDES TO ACROCARPS

QUICK GUIDES TO PLEUROCARPS

QUICK GUIDES TO SPHAGNUM

ACROCARPS

PLEUROCARPS

SPHAGNUM

LEAVES WIDE SPREADING OR RECURVED

DICRANELLA, p. 70   HEDWIGIA, p. 85   MEESIA, p. 86   PALUDELLA, p. 91   PLAGIOPUS, p. 94

LEAVES WITH BORDERS

ATRICHUM, p. 62   BRYUM, p. 65   CYRTOMNIUM, p. 69   FISSIDENS, p. 81

LEAVES WITH BORDERS

MNIUM, p. 87   PLAGIOMNIUM, p. 92   POGONATUM, p. 95   RHIZOMNIUM, p. 103

LEAVES STRAP LIKE, WITH BLUNT TIPS

BARBULA, p. 64   DIPHYSCIUM FOLIOSUM, p. 76   GRIMMIA UNICOLOR, p. 84   MEESIA ULIGINOSA, p. 87

DISTINCTIVE LEAVES II. Squarrose leaves stick out at a wide angle; re-curved leaves arch outwards and down from an erect base. Bordered leaves have elongate cells along their edges. In *Atrichum* and *Bryum* the borders may be thin and only visible with a strong lens or micro-scope. The straplike leaves here have to be blunt tipped. If they have sharp tips they go with the ordinaries on pages 34 and 35.

Within this group, most of the genera can be identified on sight. Note, as aids, the clasping bases of *Dicranella* and *Meesia*; the 3-rowed leaves of *Plagiopus*; the blue-green color of *Pogonatum* and *Cyrtomnium*, the dark green of *Grimmia* and *Diphyscium*, and the light yellow-green of *Barbula*; the double teeth of *Mnium*; the ripply leaves of *Atrichum*; and the red peristomes of the tiny *Fissidens bryoides*.

AMPHIDIUM, p. 60

BARBULA, p. 64

BRYOERYTHROPHYLLUM, p. 65

CYNODONTIUM, p. 69

DICHODONTIUM, p. 69

DICRANUM, p. 71

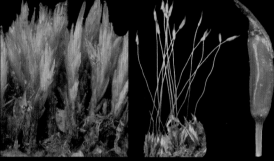

DITRICHUM, p. 77

GYMNOSTOMUM, p. 84

HYMENOSTYLIUM, p. 85

RHABDOWEISIA, p. 102

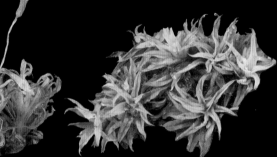

TORTELLA, p. 107

TREMATODON, p. 109

TRICHOSTOMUM, p. 109

WEISSIA, p. 111

ORDINARY ACROCARPS BY LEAF SHAPE. After the genera with distinctive leaves and shoots are accounted for, there are about 35 acrocarp genera left. We can identify most fertile plants to genus in the field. Sterile plants are much harder. To tell, say, sterile *Amphidium* from *Dichodontium*, or sterile *Ditrichum* from *Dicranella*, we need a microscope.

To sort the ordinary acrocarps we use leaf shape. The genera on this page all have long, narrow leaves, either straight-sided (linear) or tapering (lanceolate). They are narrower than the leaves on page 34, and have shorter tips than the ones on page 35. *Dicranum* may have brood branches, *Bryoerythrophyllum* has brick-red lower leaves, *Tortella* has curved leaves, *Trichostomum* broken ones. For the others, you need a microscope, capsules if you can get them, and lots of practice. In particular, *Amphidium*, *Cynodontium*, *Dichodontium*, *Gymnostomum* and *Hymenostylium* are very similar, and always require the microscope.

QUICK GUIDES TO HABITATS

QUICK GUIDES TO ACROCARPS

QUICK GUIDES TO PLEUROCARPS

QUICK GUIDES TO SPHAGNUM

ACROCARPS

PLEUROCARPS

SPHAGNUM

ANDREAEA, p. 60

AULACOMNIUM, p. 63

BRYUM, p. 65

CERATODON, p. 68

DICRANUM, p. 73

DIDYMODON, p. 76

DRUMMONDIA, p. 78

LEUCOBRYUM, p. 86

ORTHOTRICHUM, p. 89

POHLIA, p. 96

RACOMITRIUM, p. 101

SCHISTIDIUM, p. 104

TIMMIA, p. 106

TORTELLA, p. 107

ULOTA, p. 110

ZYGODON, p. 111

ORDINARY LANCEOLATE LEAVES—tapered, several times as long as wide, sharp-pointed but without an extended tip or needle tip—are found in relatively few acrocarps. Most of the sixteen we show here are field identifiable, especially when they have capsules. They are also, excepting *Didymodon* and *Zygodon*, common plants and worth knowing. Habitat helps a lot: consult the species accounts and moss maps for details.

Some useful characters are the dark color and odd capsules of *Andreaea*; the purple-red setae of *Ceratodon*; the clumped habit of *Orthotrichum* and *Ulota* and the creeping habit of *Drummondia* and *Zygodon*; the short side branches and long capsules in *Racomitrium*; the immersed capsules and long perichaetial leaves of *Schistidium*; the surface iridescence of *Pohlia*; the thick, white leaves of *Leucobryum*; and the way the leaves *Timmia* roll into tubes when they dry.

ORDINARY ACROCARPS WITH SLENDER LONG-TIPPED LEAVES*

ARCTOA, p. 61

BARTRAMIA, p. 63

BLINDIA, p. 65

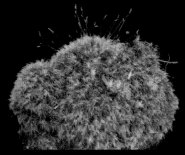

DICRANELLA, p. 70

DICRANUM, p. 71

DITRICHUM, p. 77

ONCOPHORUS, p. 89

PARALEUCOBRYUM, p. 91

PHILONOTIS, p. 92

PLEURIDIUM, p. 95

TORTELLA, p. 108

TREMATODON, p. 109

ORDINARY ACROCARPS WITH OVAL OR OBOVAL LEAVES (NOT IN TWO ROWS)

APHANORRHEGMA, p. 61

ATRICHUM, p. 62

AULACOMNIUM, p. 63

BRYUM, p. 61

IN THE LONG-TIPPED GROUP* the costa of the leaf is prolonged (excurrent) into a slender tip that is the same color and texture as the leaf below. This is a common northern-forest leaf type. The larger species are recognizable on sight. Useful characters are the pale color of *Bartramia*; the curved leaves of *Dicranum* and *Paraleucobryum* and the corkscrew ones of *Oncophorus*; and the opaque, papillose ones of *Tortella* and *Philonotis*. The small ones—*Arctoa*, *Leptobryum*, *Ditrichum*, *Dicranella*, and *Trematodon*—are virtually identical, even under a microscope. Get them with capsules or leave them be.

*Uncommon genera, not shown, that also fall here are *Cynodontium*, on cold wet ledges, *Dicranodontium*, on shaded boulders, and *Leptobryum*, on disturbed soil.

QUICK GUIDES TO HABITATS

QUICK GUIDES TO ACROCARPS

QUICK GUIDES TO PLEUROCARPS

QUICK GUIDES TO SPHAGNUM

ACROCARPS

PLEUROCARPS

SPHAGNUM

HYOPHILA, p. 85

MNIUM, p. 87

ORTHOTRICHUM, p. 89

ENCALYPTA, p. 78

PSEUDOBRYUM, p. 100

RACOMITRIUM, p. 100

RHODOBRYUM, p. 103

TETRAPHIS, p. 107

ACROCARPS WITH VEGETATIVE GEMMAE, BROOD BRANCHES, OR LEAVES WHOSE TIPS BREAK OFF

GEMMAE OR BROOD BRANCHES

gemmae

brood branch

gemmae

AULACOMNIUM, p. 63

DICRANUM FLAGELLARE, p. 71

ENCALYPTA, p. 78

POHLIA, p. 96

ORTHOTRICHUM, p. 89

GEMMAE OR BROOD BRANCHES

gemmae cup

LEAF TIPS BREAK OFF

TETRAPHIS, p. 107

ULOTA, p. 110

DICRANUM, p. 71

TRICHOSTOMUM, p. 109

TORTELLA, p. 108

THE OVAL-OBOVAL GROUP includes genera whose leaves are narrowly or broadly oval, and not in one of the distinctive groups on pages 30-32. Thus *Cyrtomnium*, although it has an oval leaf, is found in the two-rowed group (p. 30) and the bordered group (p. 32) not here.

THE VEGETATIVE REPRODUCTION group includes the genera with gemmae, brood branches and deciduous leaf tips. These are excellent field marks: the broken tips on *Dicranum viride*, for example, are the quickest way to separate it from several similar species.

ACROCARP CAPSULES

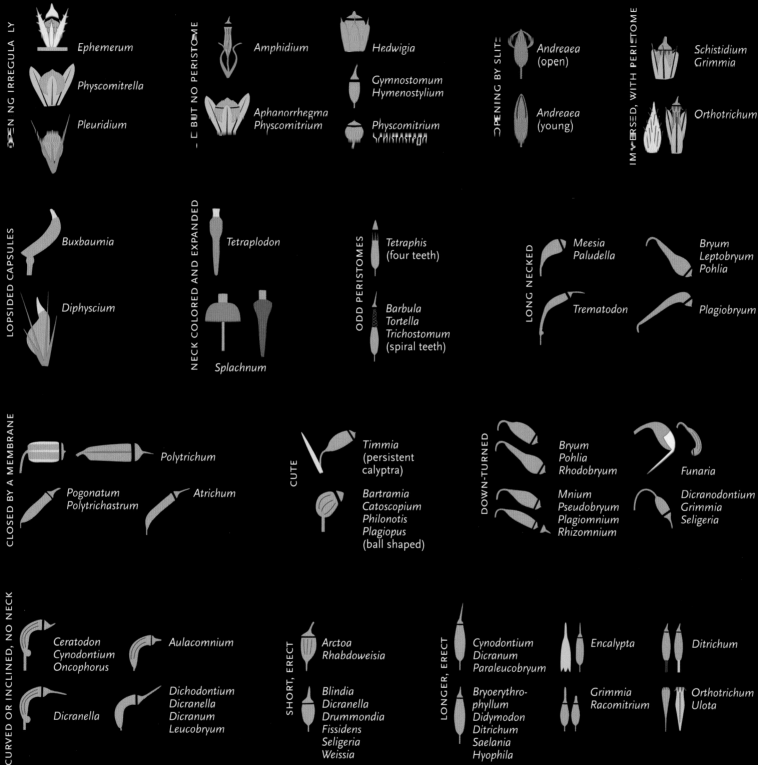

OPENING IRREGULARLY
- Ephemerum
- Physcomitrella
- Pleuridium

LID BUT NO PERISTOME
- Amphidium
- Aphanorrhegma Physcomitrium
- Hedwigia
- Gymnostomum Hymenostylium
- Physcomitrium Schistostega

OPENING BY SLIT
- Andreaea (open)
- Andreaea (young)

IMMERSED, WITH PERISTOME
- Schistidium Grimmia
- Orthotrichum

LOPSIDED CAPSULES
- Buxbaumia
- Diphyscium

NECK COLORED AND EXPANDED
- Tetraplodon
- Splachnum

ODD PERISTOMES
- Tetraphis (four teeth)
- Barbula Tortella Trichostomum (spiral teeth)

LONG NECKED
- Meesia Paludella
- Trematodon
- Bryum Leptobryum Pohlia
- Plagiobryum

CLOSED BY A MEMBRANE
- Polytrichum
- Pogonatum Polytrichastrum
- Atrichum

CUTE
- Timmia (persistent calyptra)
- Bartramia Catoscopium Philonotis Plagiopus (ball shaped)

DOWN-TURNED
- Bryum Pohlia Rhodobryum
- Mnium Pseudobryum Plagiomnium Rhizomnium
- Funaria
- Dicranodontium Grimmia Seligeria

CURVED OR INCLINED, NO NECK
- Ceratodon Cynodontium Oncophorus
- Dicranella
- Aulacomnium
- Dichodontium Dicranella Dicranum Leucobryum

SHORT, ERECT
- Arctoa Rhabdoweisia
- Blindia Dicranella Drummondia Fissidens Seligeria Weissia

LONGER, ERECT
- Cynodontium Dicranum Paraleucobryum
- Bryoerythro-phyllum Didymodon Ditrichum Saelania Hyophila
- Encalypta
- Grimmia Racomitrium
- Ditrichum
- Orthotrichum Ulota

IMPORTANT FEATURES OF CAPSULES are the shape, the length of the seta, the presence or absence of a beak, and the presence or absence of a peristome. They are helpful in almost all the genera and essential in some of the tiny ones, particularly *Dicranella*, *Ditrichum*, and the ephemerals. They are also very useful for the field identification of *Gymnostomum*, *Didymodon*, and several other genera of small, difficult, papillose mosses. Capsule types that are worth knowing include the immersed (surrounded by leaves) capsules of *Schistidium*, *Hedwigia*, *Orthotrichum*, and many ephemerals; the cylindrical capsules, closed by a drumlike membrane, of *Polytrichum* and its relatives; the asymmetrical capsules of *Buxbaumia* and *Diphyscium*; the enlarged, fragrant capsules of the dung mosses; the long beaked capsules, often curved, of *Dicranum* and its relatives; and the long-necked or down-curved capsules of *Mnium*, *Bryum*, *Meesia*, and their relatives.

QUICK GUIDES TO HABITATS

QUICK GUIDES TO ACROCARPS

QUICK GUIDES TO PLEUROCARPS

QUICK GUIDES TO SPHAGNUM

ACROCARPS

PLEUROCARPS

SPHAGNUM

37

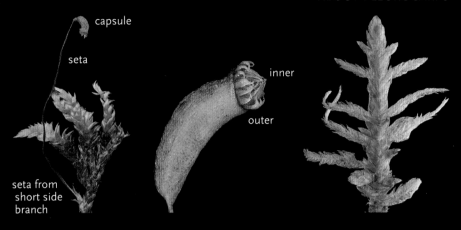

seta

capsule

seta from short side branch

inner

outer

COMMON FEATURES OF PLEUROCARPS. From left: gametophyte and sporophyte of *Hygrohypnum eugyrium*, showing irregular branching and the seta originating from a short side branch; capsule of *Hygroamblystegium tenax* with the peristome teeth in two circles; and flattened shoot of *Entodon cladorrhizans*, showing pinnate branching.

PLEUROCARPS have their sex organs on short side branches and capsules with double peristomes. Most are abundantly, and often regularly, branched. They are a natural group, thought to be about 200 million years old. They reach their greatest abundance in the understories of moist forests and permanent wetlands, often forming the deep loose colonies we describe as thick mats, thatches, fringes, arching fronds, and groves.

We have about 70 pleurocarp genera in the northern forest. With good material and a good eye, 50 to 60 can be identified with reasonable confidence in the field. Habitat, branching pattern, and leaf shape are crucial. Capsules are always helpful. Leaf details are important when you can see them, but many pleurocarps have tiny leaves that are hard to see with a lens. We concentrate on the distinctive ones in the field and bag the rest and bring them to the microscope.

SHOOTS OF PLEUROCARPS, showing distinctive branching patterns and leaf arrangements. From left: narrow, slenderly pinnate shoot of *Abietinella abietina*; frondose (flattened and fernlike), twice-pinnate shoot of *Thamnobryum alleghaniense*; and the treelike shoot with stiff, widely spreading leaves of

Rhytidiadelphus triquetrus. Above, long-pointed leaves and flattened shoot of *Taxiphyllum deplanatum*; wormlike (cylindrical) shoot with tightly shingled leaves of *Pseudocalliergon trifarium*; and the pinnate shoot with slender, erect, sharp-pointed leaves of *Tomentypnum nitens*.

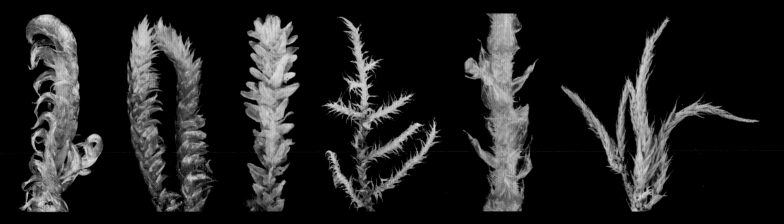

PLEUROCARP LEAVES and leaf arrangements. From left: *Hamatocaulis vernicosus*, with leaves strongly curled (falcate) and turned to one side of the branch (secund); *Hygrohypnum eugyrium*, concave, oval, falcate-secund leaves; *Anomodon minor*, blunt, oblong leaves pointing in all directions.

Above, *Campylium stellatum*, pinnately branched shoots and widely spreading (squarrose) leaves with needle points; *Hylocomium umbratum*, large stem leaves with paraphyllia (delicate fuzzy things) in between them*; and *Brachythecium populeum*, an irregularly branched shoot with long-pointed leaves.

* Paraphyllia are skinny, photosynthetic outgrowths of the stem, one or more cells wide, often threadlike and highly branched. They are sometimes described as leaflike; this is a stretch.

ANOMODON, p. 113

BRACHYTHECIUM, p. 115

BRYHNIA, p. 118

CALLICLADIUM, p. 119

CLIMACIUM, p. 122

ENTODON, p. 125

HYLOCOMIUM, p. 132

HYPNUM, p. 133

PLAGIOTHECIUM, p. 138

PLEUROZIUM, p. 141

PTILIUM, p. 143

RHYTIDIADELPHUS, p. 145

THAMNOBRYUM, p. 147

THUIDIUM, p. 148

DOMINANT UPLAND PLEUROCARPS. These are the commonest mosses of forest floors, ledges, tree bases and logs. They are large, mostly pinnate, frondose, or treelike, and highly competitive. If you are in a moist forest, much of the pleurocarp cover will be these species. Most can be recognized on sight. Branching pattern and leaf shape are the best characters. See the guides on the following pages for details.

Several are harder. *Brachythecium*, *Bryhnia*, and *Callicladium*, all common, will look ordinary at first. With practice, they will come to look ordinary in a distinctive way. *Plagiothecium* will always be hard: the species are variable and it has close relatives. The relatives tend to be scarcer; as a rule of thumb, a moss that looks like *Plagiothecium* and covers large areas probably is *Plagiothecium*.

QUICK GUIDES TO HABITATS

QUICK GUIDES TO ACROCARPS

QUICK GUIDES TO PLEUROCARPS

QUICK GUIDES TO SPHAGNUM

ACROCARPS

PLEUROCARPS

SPHAGNUM

WETLANDS

Brachythecium, p. 116 *Bryhnia*, p. 118 *Calliergon*, p. 119 *Calliergonella*, p. 120 *Campylium*, p. 120

WETLANDS

Hamatocaulis, p. 127 *Hypnum*, p. 133 *Limprichtia*, p. 136 *Rhytidiadelphus*, p. 145 *Scorpidium*, p. 146

ROCKY STREAMS

Brachythecium, p. 115 *Hygroamblystegium*, p. 130 *Hygrohypnum*, p. 130 *Sematophyllum*, p. 147 *Torrentaria*, p. 149

TREELIKE PLEUROCARPS WITH A STEM AND SPREADING BRANCHES

TREELIKE

Bryhnia, p. 148 *Brachythecium*, p. 116 *Climacium* p. 122 *Rhytidiadelphus*, p. 145

DOMINANT WETLAND PLEUROCARPS. Pleurocarps are dominant or co-dominants in rich fens, minerotrophic swamps, and in the channels and in the banks of rocky streams. They are all but absent from acid bogs and spruce swamps. The genera above are common but not easy. Be aware that wetland plants are variable; that many other, less common, genera can be mixed in; that *Limprichtia* cannot be distinguished from *Hamatocaulis* or *Hygrohypnum* from *Sematophyllum* in the field; and that *Calliergon* and *Calliergonella* can look very similar

FRONDOSE SHOOTS*

HYLOCOMIUM, p. 132

HYLOCOMIUM, p. 132

HYLOCOMIUM, p. 132

HELODIUM, p. 127

FRONDOSE SHOOTS

THAMNOBRYUM, p. 147

ISOTHECIUM, p. 135

THUIDIUM, p. 148

PTILIUM, p. 143

UPLAND, FALCATE SECUND, NO COSTA

BROTHERELLA, p. 117

HYPNUM, p. 133

CTENIDIUM, p. 123

PTILIUM, p. 143

UPLAND, FALCATE SECUND, WITH COSTA

SANIONIA, p. 145

RHYTIDIUM, p. 145

PINNATE PLEUROCARPS I. Almost half of the pleurocarp genera are pinnately branched. We subdivide and summarize them on the next three pages. Above, two upland groups. The frondose genera make flat, complex, multiply pinnate shoots, typically arching and often covering large areas. All are distinctive and easily identified. *Hylocomium splendens*, *Ptilium*, *Thamnobryum*, and *Thuidium* are very common. *Helodium* is locally common in limy fens and meadows. *Isothecium* and the other

Hylocomiums are rarer. The falcate-secund genera have long curved leaves that turn one way, like grass in a wind. *Ptilium* (boreal forest) and *Rhytidium* (limy ledges) are distinctive. The other four can be very similar. We guess at them and then confirm in the lab. *Hypnum* is the neatest, *Ctenidium* the wildest, *Sanionia* costate (if you can see it) and very curly, and *Brotherella* shiny and frizzy. But that said, all four are pinnate, and curly, and you will need to take care.

* A frondose shoot is one that looks some like a fern, or like a hand with spreading fingers: a flattened, regular or irregular spray of branches. *Ptilium* is both falcate secund and frondose.

QUICK GUIDES TO HABITATS

QUICK GUIDES TO ACROCARPS

QUICK GUIDES TO PLEUROCARPS

QUICK GUIDES TO SPHAGNUM

ACROCARPS

PLEUROCARPS

SPHAGNUM

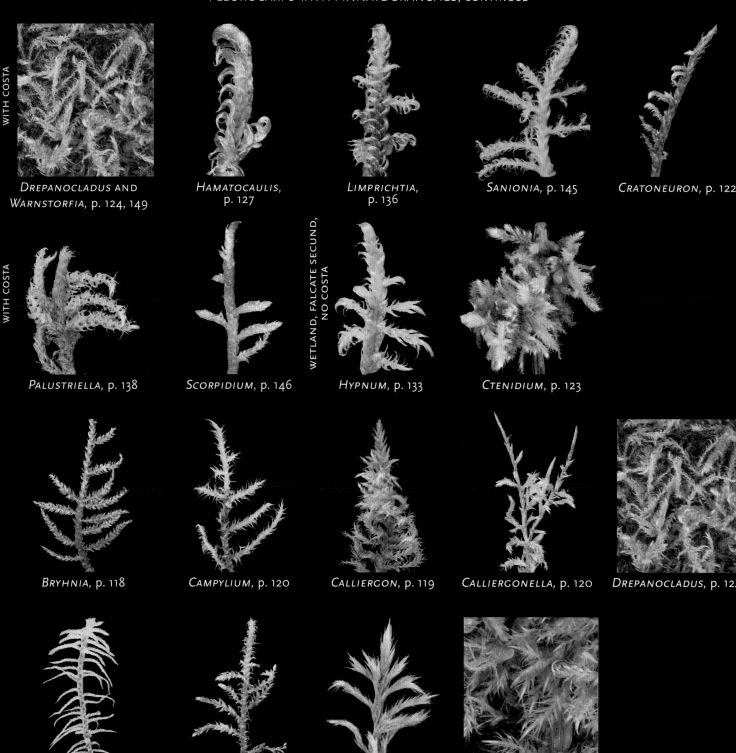

WETLAND, FALCATE SECUND, WITH COSTA

DREPANOCLADUS AND
WARNSTORFIA, p. 124, 149

HAMATOCAULIS,
p. 127

LIMPRICHTIA,
p. 136

SANIONIA, p. 145

CRATONEURON, p. 122

WETLAND, FALCATE SECUND, WITH COSTA

PALUSTRIELLA, p. 138

SCORPIDIUM, p. 146

WETLAND, FALCATE SECUND, NO COSTA

HYPNUM, p. 133

CTENIDIUM, p. 123

WETLAND, NOT FALCATE SECUND

BRYHNIA, p. 118

CAMPYLIUM, p. 120

CALLIERGON, p. 119

CALLIERGONELLA, p. 120

DREPANOCLADUS, p. 124

WETLAND, NOT FALCATE SECUND

HELODIUM, p. 127

RHYTIDIADELPHUS, p. 145

TOMENTYPNUM, p. 149

WARNSTORFIA, p. 149

PINNATE PLEUROCARPS II. The remaining falcate-secund pleurocarps (upper two rows) are wetland species. They are closely related and hard to separate in the field. *Scorpidium* has cylindrical, wormlike shoots; *Sanionia* has pleated leaves; *Cratoneuron* and *Ctenidium* have broad stem leaves; *Hamatocaulis* and *Limprichtia* are often shiny red or gold. All these characters work for some plants on some days. On others, only the microscope knows.

THE PINNATE PLEUROCARPS WITHOUT FALCATE-SECUND LEAVES are fairly easy to recognize. Note the tiny hair tips on *Cirriphyllum* and the twisted ones on *Bryhnia*; the sharp tips on *Tomentypnum* and the hooked ones on *Rhytidiadelphus*; the broad oval leaves of *Calliergon* and *Calliergonella,* and the triangular-oval stem leaves of *Helodium.* *Warnstorfia* is variable, with straight-leaved and falcate-leaved forms: microscopic confirmation is essential.

UPLAND, SPARSE, WIRY, PAPILLOSE

ABIETINELLA, p. 112

CYRTO-HYPNUM, p. 123

HAPLOCLADIUM, p. 127

HETEROCLADIUM, p. 129

RAUIELLA, p. 144

UPLAND NOT FALCATE SECUND

CALLICLADIUM, p. 119

CIRRIPHYLLUM, p. 121

ENTODON, p. 125

PLEUROZIUM, p. 141

PLEUROCARPS WITH FLATTENED SHOOTS, LEAVES NOT FALCATE SECUND

CALLICLADIUM, p. 119

ENTODON, p. 125

HOMALIA, p. 129

ISOPTERYGIOPSIS, p. 134

PLAGIOTHECIUM, p. 139

NECKERA, p. 138

PSEUDOTAXIPHYLLUM, p. 142

RHYNCHOSTEGIUM, p. 144

TAXIPHYLLUM, p. 147

THE SPARSE, WIRY, PINNATE GENERA are relatives of *Thuidium* and, like it, have papillose leaves that are dull and opaque when dry, triangular-oval stem leaves, and abundant paraphyllia along the stem. They differ from *Thuidium* in their less intricate branching. *Abietinella* is big and has strongly tapered branches. The rest are similar; we make presumptive identifications by size and substrate and then take them, quickly, to the microscope.

THE FLATTENED-SHOOT PLEUROCARPS are easy when, like *Homalia*, *Neckera*, and some *Plagiothecium*, they are strongly flattened. They are much harder when they are only slightly flattened. *Isopterygiopsis* is narrow and tiny. *Callicladium* is common and often confusing. *Plagiothecium*, *Pseudotaxiphyllum*, and *Taxiphyllum* are close. The brood bodies will pick out *Pseudotaxiphyllum*. For the others I ask Sue.

QUICK GUIDES TO HABITATS

QUICK GUIDES TO ACROCARPS

QUICK GUIDES TO PLEUROCARPS

QUICK GUIDES TO SPHAGNUM

ACROCARPS

PLEUROCARPS

SPHAGNUM

UPLAND SPECIES

Anomodon, p. 113

Bryoandersonia, p. 118

Entodon, p. 125

Myurella, p. 137

WETLAND SPECIES

Thelia, p. 148

Calliergon, p. 119

Pseudocalliergon, p. 141

Scorpidium, p. 146

SHAGGY PLEUROCARPS MAKING FRINGES AND COAT HOOKS ON THE SIDES OF TREES AND ROCKS

Anomodon, p. 113

Brachythecium, p. 115

Hylocomium, p. 132

Hypnum, p. 133

Isothecium, p. 135

Forsstroemia, p. 126

Leucodon, p. 136

Neckera, p. 138

THE WORM-BRANCHED PLEUROCARPS have cylindrical shoots with tightly overlapping leaves. The fringes and coat hooks hang off the sides of trees and rocks, sometimes dangling, sometimes curling up at their tips. All the genera are distinctive: note, for example, the dull papillose leaves of *Anomodon* and *Thelia*; the falcate-secund leaf tips of *Hypnum* and *Scorpidium* and the slender straight ones of *Brachythecium* and *Leucodon*; the broad-oval leaves of *Bryoandersonia*, *Calliergon*, *Pseudocalliergon*, and *Myurella*; and the curved branches of *Pylaisia*.

SHAGGY PLEUROCARPS, AND A LIVERWORT, MAKING FRINGES FROM THE SIDES OF TREES AND ROCKS

PLAGIOTHECIUM, p. 139

PORELLA (A LIVERWORT)

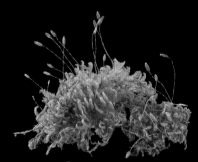

PYLAISIA, p. 143

THAMNOBRYUM, p. 147

SMALL STRINGY PLEUROCARPS MAKING TRACERIES OR THIN MATS

AMBLYSTEGIUM, p. 112

ANOMODON p. 114

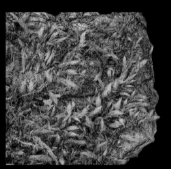

HOMOMALLIUM, p. 129

HYGROAMBLYSTEGIUM, p. 130

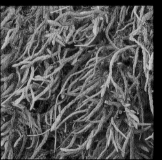

LESKEA, p. 135

LESKEELLA, p. 136

LINDBERGIA, p. 137

MYURELLA, p. 137

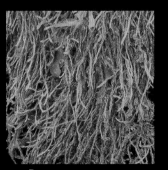

PLATYDICTYA, p. 140

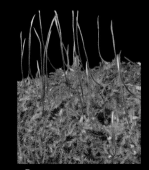

PLATYGYRIUM, p. 140

PTERIGYNANDRUM, p. 143

SCHWETSCHKEOPSIS, p. 146

THE SMALL STRINGIES are species of bark, rock, and logs that branch irregularly and either make thin flat mats or traceries of individual stems. They are hard to identify without a microscope and not all that easy with one. For field clues, note the shiny leaves of *Homomallium* and the dull ones of *Leskea* and *Pterigynandrum*; the brood branches of *Platygyrium* and *Leskeella*; the wide-spreading leaves of *Lindbergia*; the minute leaves, barely 0.5 mm long in *Platydictya*; and the concave, oval leaves of *Myurella* and *Pterigynandrum*.

QUICK GUIDES TO HABITATS

QUICK GUIDES TO ACROCARPS

QUICK GUIDES TO PLEUROCARPS

QUICK GUIDES TO SPHAGNUM

ACROCARPS

PLEUROCARPS

SPHAGNUM

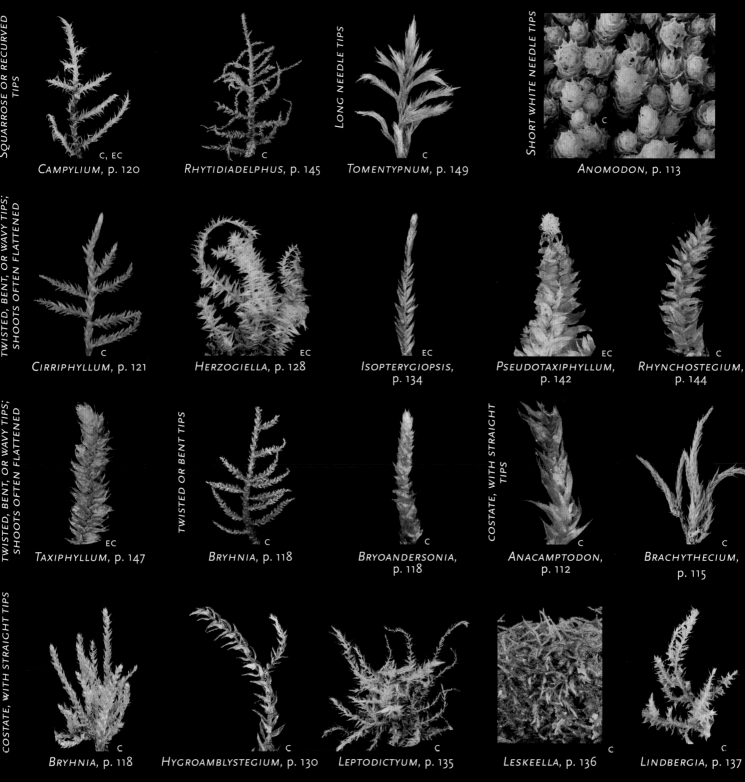

SQUARROSE OR RECURVED TIPS

C, EC
CAMPYLIUM, p. 120

C
RHYTIDIADELPHUS, p. 145

LONG NEEDLE TIPS

C
TOMENTYPNUM, p. 149

SHORT WHITE NEEDLE TIPS

C
ANOMODON, p. 113

TWISTED, BENT, OR WAVY TIPS; SHOOTS OFTEN FLATTENED

C
CIRRIPHYLLUM, p. 121

EC
HERZOGIELLA, p. 128

EC
ISOPTERYGIOPSIS, p. 134

EC
PSEUDOTAXIPHYLLUM, p. 142

C
RHYNCHOSTEGIUM, p. 144

TWISTED, BENT, OR WAVY TIPS; SHOOTS OFTEN FLATTENED

EC
TAXIPHYLLUM, p. 147

TWISTED OR BENT TIPS

C
BRYHNIA, p. 118

C
BRYOANDERSONIA, p. 118

COSTATE, WITH STRAIGHT TIPS

C
ANACAMPTODON, p. 112

C
BRACHYTHECIUM, p. 115

COSTATE, WITH STRAIGHT TIPS

C
BRYHNIA, p. 118

C
HYGROAMBLYSTEGIUM, p. 130

C
LEPTODICTYUM, p. 135

C
LESKEELLA, p. 136

C
LINDBERGIA, p. 137

THE PLEUROCARPS ON THIS PAGE have slender leaf tips that often form a short or long point. They are hard. In the lab we sort them by presence or absence of a strong costa (C, costate; EC, ecostate). In the field we pick out the squarrose leaves of *Campylium* and *Rhytidiadelphus*; the long slender tips of *Tomentypnum* and the tiny white ones of *Anomodon rostratus*; the broad, concave stem leaves of *Bryhnia* and *Bryoandersonia*; the remarkable hair tips of *Cirriphyllum*; the tiny, flat, tapering shoots of *Isopterygiopsis*; and the shiny, wide-spreading leaves and somewhat flattened shoots of *Herzogiella, Pseudotaxiphyllum, Rhynchostegium,* and *Taxiphyllum.* Beyond that, habitat helps: *Hygroamblystegium* and *Leptodictyum* are aquatics; *Bryhnia graminicolor* is on limy soil; *Leskeella* and *Lindbergia* are on tree bark; *Anacamptodon* in holes and crotches in trees; and *Brachythecium* almost anywhere a pleurocarp can grow.

QUICK GUIDES TO HABITATS

QUICK GUIDES TO ACROCARPS

QUICK GUIDES TO PLEUROCARPS

QUICK GUIDES TO SPHAGNUM

ACROCARPS

PLEUROCARPS

SPHAGNUM

BROAD-LEAVED PLEUROCARPS WITH OVAL OR OBLONG LEAVES; TIPS ROUNDED OR SHORT POINTED*

ANOMODON, p. 113

BRACHYTHECIUM, p. 115

BRYOANDERSONIA, p. 118

CALLIERGON, p. 119

CALLIERGONELLA, p. 120

EURHYNCHIUM, p. 125

FONTINALIS, p. 126

HOMALIA, p. 129

HYGROHYPNUM, p. 130

NECKERA, p. 138

PLATYLOMELLA, p. 141

PTERIGYNANDRUM, p. 143

PLEUROZIUM, p. 141

PSEUDOCALLIERGON, p. 141

SCORPIDIUM, p. 146

TORRENTARIA, p. 149

THE BROAD-LEAVED PLEUROCARPS have oval, oblong, or nearly round leaves. Some are blunt, some have small points, some have blunt tips that look pointed. None of them have long tapering points. Most are easily identifiable in the field. Key characters are the flattened shoots of *Homalia* and *Neckera*; the cylindrical ones of *Pseudocalliergon*, *Scorpidium* and *Bryoandersonia;* the oblong-oval leaves of *Anomodon*; the aquatic habitat of *Fontinalis*, *Hygrohypnum*, *Torrentaria*, and *Platylomella;*

the falcate-secund leaves of *Hygrohypnum* and *Scorpidium*; the broad, blunt leaves and red stem of *Pleurozium* in the uplands; and the blunt leaves and green stem of *Calliergon* and *Calliergonella* in wetlands.

This leaves three genera. *Pterigynandrum* is a tiny dark creeper on rocks. *Brachythecium rivulare* is a big, often treelike species of wetlands with decurrent leaves. The Eurynchiums are small and inconspicuous, often forming sparse mats on moist, partly shaded soil.

* If the leaves are closely overlapping and the shoots cylindrical, go to the worm-branched group on p. 44.

AQUATICS

DICHELYMA, p. 124

HYGROHYPNUM, p. 130

WARNSTORFIA, p. 149

WETLAND

PALUSTRIELLA, p. 138

WETLAND

SCORPIDIUM, p. 146

UPLAND

BRACHYTHECIUM VELUTINUM, p. 117

PYLAISIADELPHA, p. 144

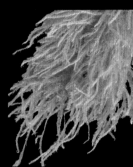

HYPNUM, p. 133

PLEUROCARPS WITH BROOD BRANCHES THAT ARE VISIBLE WITH A LENS

LESKEELLA, p. 136

LEUCODON, p. 136

PLAGIOTHECIUM, p. 139

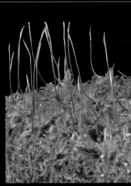

PLATYGYRIUM, p. 140

ABOVE, MOST PLEUROCARPS WITH FALCATE-SECUND LEAVES are pinnate. The ones that aren't are worth learning. *Dichelyma*, with three-ranked leaves with sharp keels, and *Scorpidium* and *Hygrohypnum* with oval concave leaves, are distinctive. Most Hypnums, pinnate or not, still look like Hypnums. *Warnstorfia* and *Palustriella* are average *Drepanocladus*-like plants that have to go to the microscope. *Brachythecium velutinum* is small, shiny, and has long-tapering leaf tips and short capsules. *Pylaisiadelpha* has nothing: Bruce Allen says it is "one of the hardest mosses to recognize in eastern North America." We agree.

ABOVE AND RIGHT, BROOD BRANCHES (small, deciduous stems) are rare in pleurocarps. Where they do occur, they are very helpful. On tree bark, they will separate *Leskeella* and *Platygyrium* from *Pylaisia*, *Amblystegium* and *Leskea*. On soil and rocks, they are the best way to separate *Pseudotaxiphyllum* and (much less commonly) *Plagiothecium* from *Taxiphyllum*.

PSEUDOTAXIPHYLLUM, p. 142

ERECT, SMALL

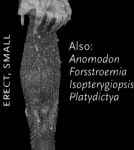

Also:
Anomodon
Forsstroemia
Isopterygiopsis
Platydictya

ANACAMPTODON, p. 112

ERECT, LARGER, SHORT BEAKS

Also:
Climacium
Homalia
Leucodon
Thamnobryum

ANOMODON, p. 113

ERECT, LARGER, NO BEAKS

Also:
Anomodon
Entodon
Leskea
Leskeella
Lindbergia
Platygyrium
Pterigynandrum
Pylaisia
Thelia

PYLAISIA, p. 143

INCLINED, SMALL

Myurella
Platydictya

Homomallium

INCLINED, LONG BEAKS

TORRENTARIA, p. 149

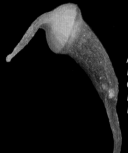

Also:
Bryoandersonia
Cirriphyllum
Eurhynchium
Rhynchostegium

SEMATOPHYLLUM, p. 147

INCLINED, SHORT BEAKS

Also:
Drepanocladus
Taxiphyllum
Thuidium

CALLICLADIUM, p. 119   PLAGIOTHECIUM,   BROTHERELLA,
p. 139   p. 117

INCLINED, SHORT, NO BEAKS

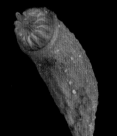

HYGROHYPNUM,
p. 130

BRACHYTHECIUM, p. 115

Also:
Bryhnia
Calliergonella
Cratoneuron
Drepanocladus
Herzogiella
Heterocladium
Hylocomium
Pleurozium
Rhytidiadelphus
Schwetschkeopsis
Scorpidium

INCLINED, LONGER, NO BEAKS

Also:
Amblystegium   *Hygrohypnum*
Calliergon   *Hypnum*
Campylium   *Isothecium*
Cratoneuron   *Leptodictyum*
Ctenidium   *Plagiothecium*
Drepanocladus   *Platylomella*
Haplocladium   *Ptilium*
Helodium   *Rhytidium*
Hygroamblystegium *Tomentypnum*

SURROUNDED BY LEAVES

Fontinalis
Neckera

Dichelyma

PLEUROCARP CAPSULES don't vary much: all have double peristomes, none are ball shaped or necked or furrowed, and only two are immersed in the perichaetial leaves. There are far more genera than there are types of capsules, and for many laboratory identifications it doesn't matter whether you have capsules or not.*

In the field, capsules come into their own. Sometimes they are all you need to see. On tree bark and logs, for example, the flaring peristome teeth of *Anacamptodon*, the short capsules of the Brachytheciums, the erect capsules of *Entodon*, and the long inclined ones of *Callicladium* are give-aways. More often, they are the easy way to identify species that would otherwise go to the microscope. Some examples: *Hypnum imponens* and *Brotherella recurvans*, separated by the beak in *Brotherella*; *Hypnum pallescens* and *Amblystegium varium*, with long capsules, vs. *Brachythecium reflexum*, with short; *Sematophyllum*, with beaked capsules, vs. *Hygrohypnum*, with unbeaked; and *Eurhynchium* and *Rhynchostegium*, with long-beaked capsules vs. *Brachythecium* and *Bryhnia*, without beaks.

* *Brachythecium* is the big exception. For *Brachythecium*, always try to get capsules.

QUICK GUIDES TO HABITATS

QUICK GUIDES TO ACROCARPS

QUICK GUIDES TO PLEUROCARPS

QUICK GUIDES TO SPHAGNUM

ACROCARPS

PLEUROCARPS

SPHAGNUM

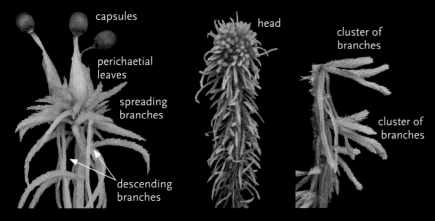

capsules

perichaetial leaves

spreading branches

descending branches

head

cluster of branches

cluster of branches

PARTS OF *SPHAGNUM*, in *Sphagnum fimbriatum* (left) and *wulfianum* (center and right). The branches are borne in clusters along the stem; each cluster has spreading and descending branches. The branches originate within the head, and elongate as the stem grows. The capsules develop within the perichaetial leaves and are raised on stalks called pseudopodia.* When ripe, they dry, shrink, and explode.

THE SPHAGNUMS are an old, unique lineage that originated in the Permian and reached its modern form in the early Jurassic. They differ from all other mosses in having clustered branches in a head and a mosaic of large water-storage cells and small photosynthetic cells in their leaves.

We treat 29 species here, leaving out some rarities and recent segregates and treating *Sphagnum reflexum* and *subsecundum* as complexes. We think of these 29 as a starter set: enough names to cover everything you are likely to see, with the option of adding more if you need them.

About 24 of these species can be recognized in the field. Color, branching, leaf arrangement, and leaf shape are key. Some field identifications will be certain, some tentative. Sphagnums vary a lot, both macroscopically and microscopically. Some fit our categories, some don't. When they don't, it's our problem, not theirs.

HEAD AND BRANCHES. From left, *Sphagnum capillifolium*, with small, dense, round heads and branches spreading in all directions; *S. warnstorfii*, flat-topped heads with the branches in vertical rows and young descending branches in a row between the spreading ones. The branches with red tips have antheridia.

Center, *S. teres*, a loosely five-parted head with a central terminal bud. Above, *S. girgensohnii* with the young descending branches (circled) in a single row; and *S. recurvum* with the young descending branches in two rows.

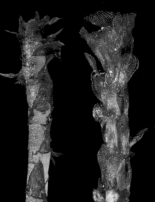

BRANCHES AND STEM LEAVES. Left group, branches from *S. centrale, compactum,* and *subsecundum*: all with deeply concave oval leaves. Middle group, branches from *S. wulfianum, warnstorfii,* and *pulchrum*: all with narrower and less concave leaves with longer tips than the first group. Right group, stems and stained stem leaves of *S. capillifolium, fimbriatum,* and *torreyanum*. The stem leaves of *capillifolium* are long rounded-triangular; of *fimbriatum*, broad and fringed; and of *torreyanum*, small and triangular.

* Pseudopodia, false-feet, because they are maternal (gametophytic) tissue, as opposed to setae, which come from the fertilized embryo and are thus sporophytic tissue.

Our 29 sphagnums are traditionally divided into 7 sections. The largest are *Acutifolia* (11 species), *Cuspidata* (7 species). and *Sphagnum* (5 species). *Subsecunda*, treated as 2 species here, contains 4 other NFR species that are close to *subsecundum*.

The sections are useful because they tell you who is related to whom. But they are also frustrating almost all the characters we use in the field occur in multiple sections. Conspicuous terminal buds, for example, occur in three sections, five-parted heads occur in four sections, and recurved leaves, for example, can occur in six sections.*

The bottom line is that the sections are a great way of organizing species you know and a poor way of learning new species. Use them by all means but only after you know the common species. When starting off, get the distinctive species and species groups done first; then do the sections.

SECTION *ACUTIFOLIA*. Common species of bogs, swamps, fens, and forests. Most species with either intense red or brown colors, small rounded heads, or broad fringed stem leaves fall here. Above, clockwise: *warnstorfii, capillifolium, fuscum, girgensohnii, quinquefarium,* and *fimbriatum.*

SECTION *CUSPIDATA*. Common species of bogs and swamps. Never red, often yellow, gold, brown, or black. Often with either long curved branch leaves, small triangular stem leaves, or young descending branches in two rows. Clockwise: *majus, cuspidatum, recurvum, riparium, pulchrum, tenellum.*

SECTION *SPHAGNUM*. Common species of swamps, bogs, and fens, recognizable in the field by their thick branches and blunt, oval, deeply concave branch leaves. The tips of the branch leaves are closed like a hood. Above, clockwise: *magellanicum, imbricatum, palustre, papillosum,* and *centrale.*

SECTIONS *POLYCLADA* AND *RIGIDA*: *Sphagnum wulfianum* (Section *Polyclada*), left, uncommon in cedar swamps, has dense round heads and six or more branches per bundle. *S. compactum* (Section *Rigida*), right, common on rock or sand, recognized by the deeply concave leaves with gravy-boat tips.

SECTIONS *SQUARROSA* AND *SUBSECUNDA*: *S. teres* (left) and *squarrosum* (center) of Section *Squarrosa* have oblong-oval stem leaves and visible terminal buds. *S. subsecundum* (right) of Section *Subsecunda* has deeply concave branch leaves with pointed tips and curved upper branches.

* With practice, many common species can be identified to section in the field. There are useful characters, for example, that will distinguish many *Acutifolia* from many *Cuspidata*. But there are no field characters that will distinguish all *Acutifolia* from all *Cuspidata*. For that, the microscope is needed.

QUICK GUIDES TO HABITATS

QUICK GUIDES TO ACROCARPS

QUICK GUIDES TO PLEUROCARPS

QUICK GUIDES TO SPHAGNUM

ACROCARPS

PLEUROCARPS

SPHAGNUM

RECURVUM GROUP, p. 155

PULCHRUM, p. 154

MAJUS, p. 154

TORREYANUM, p. 154

PALUSTRE, p. 157

MAGELLANICUM, p. 156

PAPILLOSUM, p. 157

IMBRICATUM, p. 158

THE COLORS OF *SPHAGNUM* develop in summer and are strongest in full sun. Reds and deep browns are species specific and useful for identification; yellows and pale browns are common and uninformative. Most colored species produce green forms in shade; a green *Sphagnum* can be almost anything. Reds, pinks, and deep browns are commonly Sections *Acutifolia* and *Sphagnum*. *Capillifolium* is red, pink or green; *rubellum* is typically blood-red, or pink in the shade; *russowii* is often mottled, *warnstorfii* an intense purple-red, and *magellanicum* a paler purple-red. The most intense browns occur in *fuscum* and *flavicomans* of Section *Acutifolia* and in *pylaesii* of Section *Subsecunda*. Intense orange-browns or golden browns occur in *papillosum* (Section *Sphagnum*), *pulchrum* (Section *Cuspidata*), and *teres* (Section *Squarrosa*). Dirty greens and blacks are very characteristic of the semi-aquatic species *majus* and *torreyanum* in Section *Cuspidata*.

QUICK GUIDES TO HABITATS

QUICK GUIDES TO ACROCARPS

QUICK GUIDES TO PLEUROCARPS

QUICK GUIDES TO SPHAGNUM

ACROCARPS

PLEUROCARPS

SPHAGNUM

COLORED SPHAGNUMS: RED, BROWN, YELLOW, ORANGE, AND BLACKISH

SECTION *SQUARROSA*

TERES, p. 158

SECTION *SUBSECUNDA*

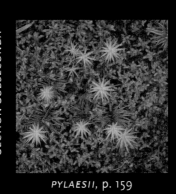

PYLAESII, p. 159

SUBSECUNDUM GROUP, p. 159

SECTION *RIGIDA*

COMPACTUM, p. 156

DOMINANT SPHAGNUMS, COVERING LARGE AREAS

SECTION *ACUTIFOLIA*

CAPILLIFOLIUM, p. 150

FLAVICOMANS, p. 151

FUSCUM, p. 151

GIRGENSOHNII, p. 152

SECTION *ACUTIFOLIA*

RUBELLUM, p. 150

SECTION *CUSPIDATA*

MAJUS, p. 154

PULCHRUM, p. 154

RECURVUM GROUP, p. 155

SECTION *CUSPIDATA*

TORREYANUM, p. 154

SECTION *SPHAGNUM*

MAGELLANICUM, p. 156

PALUSTRE, p. 157

PAPILLOSUM, p. 157

THESE SPHAGNUMS are, in our experience, the commonest species in the NFR. Excepting *magellanicum*, which can grow almost anywhere, each has a preferred range of habitats, pH, and height relative to the water table. *Capillifolium*, *flavicomans*, and *fuscum* are species of acid hummocks; *recurvum*, *papillosum*, and *rubellum* grow low on hummocks and in wet lawns in bogs; *pulchrum* is a lawn species of poor fens; *torreyanum* and *majus* are semiaquatic species of wet fens; *girgensohnii* and *palustre* are shade species, *girgensohnii* in conifer swamps and *palustre* in more fertile mixed swamps; *recurvum*, which has a wide ecological latitude, often grows with *girgensohnii* in conifer swamps.

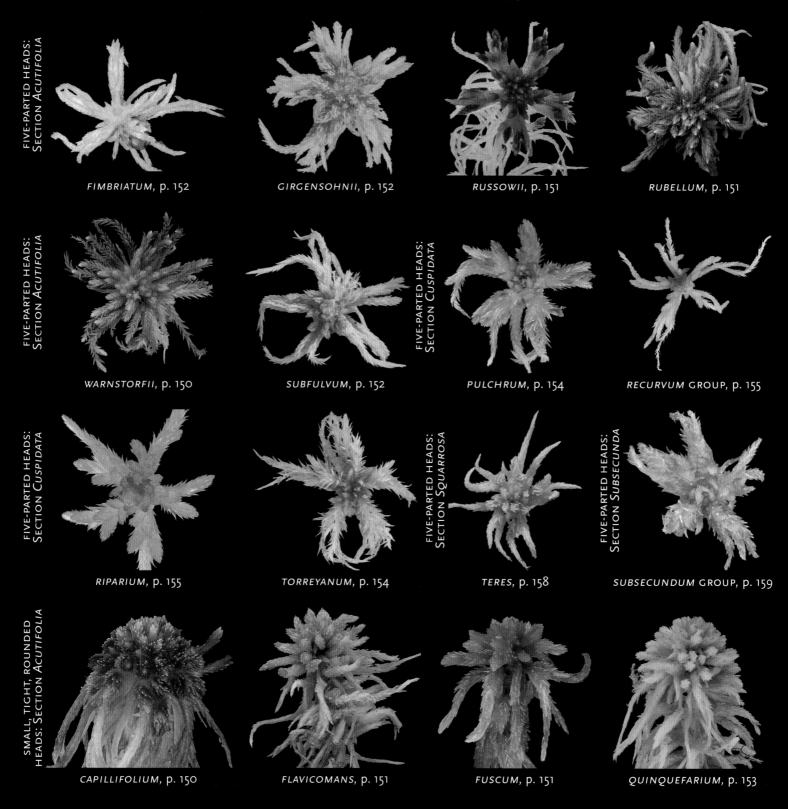

FIVE-PARTED HEADS:
SECTION *ACUTIFOLIA*

FIMBRIATUM, p. 152

GIRGENSOHNII, p. 152

RUSSOWII, p. 151

RUBELLUM, p. 151

FIVE-PARTED HEADS:
SECTION *ACUTIFOLIA*

WARNSTORFII, p. 150

SUBFULVUM, p. 152

FIVE-PARTED HEADS:
SECTION *CUSPIDATA*

PULCHRUM, p. 154

RECURVUM GROUP, p. 155

FIVE-PARTED HEADS:
SECTION *CUSPIDATA*

RIPARIUM, p. 155

TORREYANUM, p. 154

FIVE-PARTED HEADS:
SECTION *SQUARROSA*

TERES, p. 158

FIVE-PARTED HEADS:
SECTION *SUBSECUNDA*

SUBSECUNDUM GROUP, p. 159

SMALL, TIGHT, ROUNDED
HEADS: SECTION *ACUTIFOLIA*

CAPILLIFOLIUM, p. 150

FLAVICOMANS, p. 151

FUSCUM, p. 151

QUINQUEFARIUM, p. 153

FIVE-PARTED HEADS in which the branches radiate in five directions with gaps in between occur regularly in 12 species in 4 different sections and irregularly in several others. They are useful for some species pairs—*rubellum* usually has them, *capillifolium* doesn't—but are too common to be much use overall.

ROUNDED HEADS are less common but also less distinct. Use them but allow for ecological variation. They occur (above) in the hummock forming species of Section *Acutifolia*, which often make tightly packed, bumpy mounds; in *wulfianum*, page 55, which has big heads and is free standing, like little trees; and in *compactum*, page 55, in which the heads form a dense mat and are barely visible.

LARGE, ROUNDED, FREE-STANDING HEADS: SECTION *POLYCLADA*

WULFIANUM, p. 156

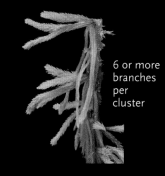

6 or more branches per cluster

SMALL, CLOSELY PACKED HEADS, MAKING A MAT: SECTION *RIGIDA*

COMPACTUM, p. 156

branch leaves with elongate channeled blunt tips

CONSPICUOUS TERMINAL BUD: SECTION *ACUTIFOLIA*

ANGERMANICUM, p. 153

FIMBRIATUM, p. 152

CONSPICUOUS TERMINAL BUD: SECTION *CUSPIDATA*

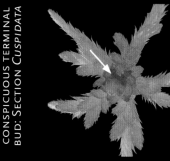

RIPARIUM, p. 155

CONSPICUOUS TERMINAL BUD: SECTION *SQUARROSA*

TERES, p. 158

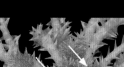

SQUARROSUM, p. 158

HEADS TINY: SECTION *SUBSECUNDA*

PYLAESII, p. 159

TERMINAL BUDS are cones of young branches and leaves, located in the center of the head. They are a great field character. Tiny ones occur in most species. Large ones occur in the five species shown and in *Sphagnum platyphyllum*, an uncommon member of the *subsecundum* complex that we haven't photographed. Really big ones that stick up are restricted to two species, *fimbriatum* and *teres*.

SPHAGNUM PYLAESII is an atypical species with tiny heads, short spreading branches, and no descending branches. The stem leaves are large and oval; the plants are small and dark brown or black. It is semiaquatic, forming thin mats where seepage flows over rocks, and floating in shallow pools in swamps.

QUICK GUIDES TO HABITATS

QUICK GUIDES TO ACROCARPS

QUICK GUIDES TO PLEUROCARPS

QUICK GUIDES TO SPHAGNUM

ACROCARPS

PLEUROCARPS

SPHAGNUM

THICK BRANCHES, BLUNT CONCAVE LEAVES: SECTION *SPHAGNUM*

CENTRALE, p. 157

MAGELLANICUM, p. 156

IMBRICATUM, p. 158

THICK BRANCHES, BLUNT CONCAVE LEAVES: SECTION *SPHAGNUM*

PAPILLOSUM, p. 157

PALUSTRE, p. 157

OVAL, CONCAVE, POINTED BRANCH LEAVES: SECTION *CUSPIDATA*

stem leaves oboval, longer than branch leaves

TENELLUM, p. 155

OVAL, CONCAVE, POINTED BRANCH LEAVES: SECTION *RIGIDA*

COMPACTUM, p. 156

OVAL, CONCAVE, POINTED BRANCH LEAVES: SECTION *SUBSECUNDA*

PYLAESII, p. 159

SUBSECUNDUM GROUP, p. 159

SPHAGNUM BRANCH LEAVES vary from spoon shaped and concave to slender and long pointed. The species on this page have deeply concave, oval branch leaves. Section *Sphagnum* (first two lines) can be recognized at sight from its short, thick branches and branch leaves with closed tips, like a hood. The species are similar, and the red forms of *magellanicum* are the only ones that can be recognized in the field.

THE REMAINING SPECIES have branch leaves that are similar to those of Section *Sphagnum* but somewhat more pointed. *Tenellum* has slender branches with well-separated leaves and stem leaves that are similar to the branch leaves but larger. *Pylaesii* is unique. *Compactum* has gravy-boat branch leaves which are pinched in to a channeled tip. *Subsecundum's* leaves are close to those of Section *Sphagnum*, but its branches are slender, tapering, and often curved.

BRANCH LEAVES LONG AND SLENDER: SECTION *CUSPIDATA*

CUSPIDATUM, p. 153

MAJUS, p. 154

TORREYANUM, p. 154

BRANCH LEAVES IN ROWS: SECTION *ACUTIFOLIA*

WARNSTORFII, p. 150

RUBELLUM, p. 150

IN ROWS: SECTION *CUSPIDATA*

PULCHRUM, p. 154

EXTRA BRANCHES: SECTION *ACUTIFOLIA*

three spreading branches

QUINQUEFARIUM, p. 153

EXTRA BRANCHES: SECTION *POLYCLADA*

6 or more branches per cluster

WULFIANUM, p. 156

SHORT FLATTENED BRANCHES: SECTION *ACUTIFOLIA*

short flattened branch

large, oval stem leaves

ANGERMANICUM, p. 153

PAIRED DESCENDING BRANCHES: SECTION *CUSPIDATA*

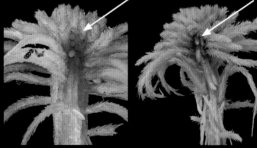

RECURVUM GROUP, p. 155

EXCEPTING ANGERMANICUM, the species on this page have longer and more pointed leaves than those on page 56. *Cuspidatum, majus,* and *torreyanum,* the semi-aquatic species of Section *Cuspidata,* have long points that are often curved and secund. *Warnstorfii* and *pulchrum* have branches in which the leaves are in lengthwise rows that extend almost to the tip. *Rubellum* and *recurvum* can be similar but not consistently. *Quinquefarium* and *wulfianum* have extra branches in the clusters. The extra branches are hard to see in the field; the recurved branch leaves of *quinquefarium* and large lollipop heads of *wulfianum* are good field characters. *Angermanicum* has small, stubby, flattened branches in the head, combined with very large stem leaves. The *recurvum* group has the young descending branches in pairs, and tiny triangular stem leaves that often point downwards. You don't see either all the time, but when you do they are definitive.

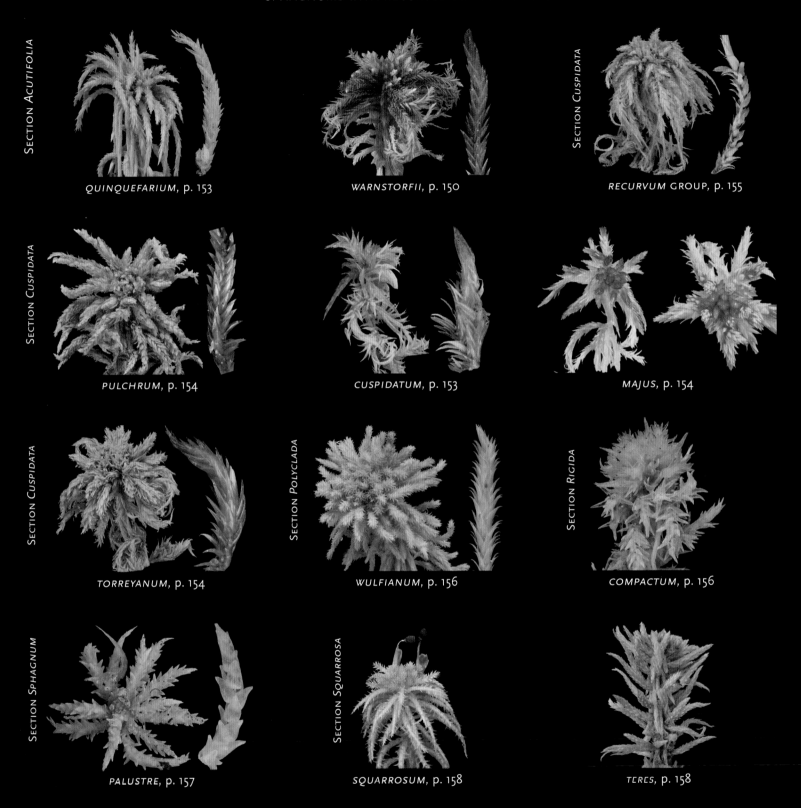

SECTION *ACUTIFOLIA*

QUINQUEFARIUM, p. 153

WARNSTORFII, p. 150

SECTION *CUSPIDATA*

RECURVUM GROUP, p. 155

SECTION *CUSPIDATA*

PULCHRUM, p. 154

CUSPIDATUM, p. 153

MAJUS, p. 154

SECTION *CUSPIDATA*

TORREYANUM, p. 154

SECTION *POLYCLADA*

WULFIANUM, p. 156

SECTION *RIGIDA*

COMPACTUM, p. 156

SECTION *SPHAGNUM*

PALUSTRE, p. 157

SECTION *SQUARROSA*

SQUARROSUM, p. 158

TERES, p. 158

RECURVED LEAVES, which arch outwards or backwards from an erect base, occur in six of our seven sections of *Sphagnum*. They are best seen in dry plants. They are very useful for some pairs of otherwise confusing species. Examples include: separating *quinquefarium* and *warnstorfii* from their look-alikes *capillifolium* and *rubellum*; separating the *recurvum* group from *girgensohnii*; in combination with the long needle-tipped leaves, separating *cuspidatum*, *majus*, and *torreyanum* from everything else; separating mats of *compactum* from *capillifolium* in the boreal forest; and picking out *squarrosum* in wooded swamps.

Two details are worth noting. First, the leaves of the *recurvum* group, which are inrolled when wet, unroll and flatten out when they dry. So far as we know, nothing else does this. And second: so far as we know, only *squarrosum* and some forms of *palustre* are consistently recurved when wet.

QUICK GUIDES TO HABITATS

QUICK GUIDES TO ACROCARPS

QUICK GUIDES TO PLEUROCARPS

QUICK GUIDES TO SPHAGNUM

ACROCARPS

PLEUROCARPS

SPHAGNUM

SPHAGNUMS WITH DISTINCTIVE STEM LEAVES

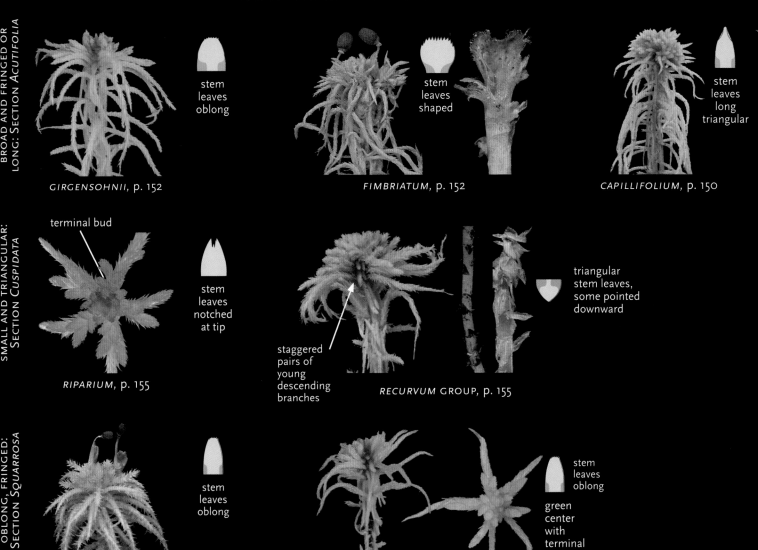

BROAD AND FRINGED OR LONG: SECTION *ACUTIFOLIA*

stem leaves oblong

GIRGENSOHNII, p. 152

stem leaves shaped

FIMBRIATUM, p. 152

stem leaves long triangular

CAPILLIFOLIUM, p. 150

SMALL AND TRIANGULAR: SECTION *CUSPIDATA*

terminal bud

stem leaves notched at tip

RIPARIUM, p. 155

staggered pairs of young descending branches

triangular stem leaves, some pointed downward

RECURVUM GROUP, p. 155

OBLONG, FRINGED: SECTION *SQUARROSA*

stem leaves oblong

SQUARROSUM, p. 158

stem leaves oblong

green center with terminal bud

TERES, p. 158

TWO DIFFICULT SPECIES: BROWNISH, WEAKLY FIVE-PARTED HEADS, BROWN STEMS

SECTION *ACUTIFOLIA*

SUBFULVUM, p. 152

SECTION *SUBSECUNDA*

SUBSECUNDUM GROUP, p. 159

STEM LEAVES are much used in *Sphagnum* microscopy but are of limited use in routine field work. We illustrate only the most distinctive: fan shaped and fringed in *fimbriatum*; oblong and fringed in *girgensohnii*, oval-oblong in Section *Squarrosa*; short triangular in several *Cuspidata*, long triangular in *capillifolium*. The last is especially useful for the small, green woodland plants that might be *capillifolium* or might not.

THE SUB-SPHAGNUMS. Two species that often give trouble are *subfulvum* and *subsecundum*. Both are average-looking, brownish Sphagnums with five-parted heads and dark stems. *Subfulvum* looks like a shade form of *fuscum; subsecundum,* a complex of several intergrading species, looks a bit like a stringy form of something from Section *Sphagnum.* Often we recognize them by eliminating everything else.

59

AMPHIDIUM [U] ⛢ Slender-leaved mosses in dense tufts, with a thatch of several years of old leaves at their base

slender leaves with recurved edges and papillae that cross cell walls; rocks with seepage, often by water

TWO SPECIES OF SMALL MOSSES that form dense cushions or hanging mossicles on wet rocks. One or the other is locally common in the spray zone of brooks and on ledges with limy seepage. Their marks are the long, parallel-sided leaves with recurved edges, opaque when dry, with papillae that cross the walls of the cell. *Lapponicum* is found throughout the NFR; *mougeotii* is more eastern. They look the same and grow in similar habitats; we separate them by the shape of the papillae. *Gymnostomum* and *Dichodontium* look similar but are half the size. *Hymenostylium* is very similar and needs the scope.

AMPHIDIUM MOUGEOTII

ANDREAEA [D] [F] [M] Small dark mosses with incurved concave leaves, thick-walled cells, and capsules opening by vertical slits

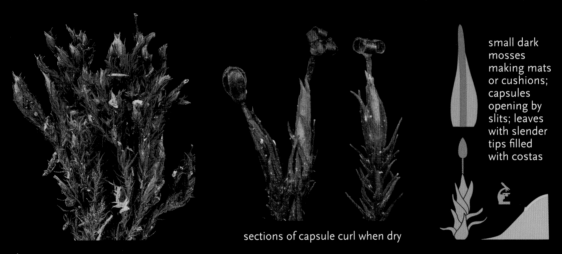

small dark mosses making mats or cushions; capsules opening by slits; leaves with slender tips filled with costas

sections of capsule curl when dry

A NORTHERN GENUS of small black or red mosses, common on wet or intermittently wet cliffs, often forming large patches in water tracks. Also on rocks in the splash-zone of brooks and on boulders in lakes. Leaves small, with thick-walled cells. Colors vary from black to dark red or green. The capsules, which open by four slits, are unique. Compare *Grimmia* and *Schistidium*, also dark and common on rocks but never red, and often with needle tips.

ANDREAEA ROTHII

leaves without costas, broader at tip

BOTH SPECIES OF *ANDRE-AEA* are pioneers on exposed rock faces. *Rupestris*, with short leaves without costas, is the most widespread in the NFR and the one most often seen on exposed rocks in the high mountains. *Rothii*, with longer leaf tips filled with the costa, is restricted to the eastern NFR. It is commonest near water and on wet cliffs with seepage.

ANDREAEA RUPESTRIS＊

＊ The photographs, on this page and elsewhere, are brightly lit for sharpness and detail. Many species are much darker in the field.

ANOMOBRYUM JULACEUM R Small plants with shiny cylindrical shoots tapering to pointed tips; leaves tightly shingled

cylindrical shoots; tightly overlapping oval leaves with long cells and no borders; crevices in rocks with limy seepage

A RARE NORTHERN SPECIES, found among other bryos in crevices of cold cliffs, often with limy seepage. The pale color and slender tapering shoots are good field characters. Compare *Plagiobryum* and *Bryum argenteum*, with sharp-tipped leaves and shorter cells; and *Myurella julacea*, with rounder leaves, short cells, and creeping basal stems which form a mat.

ANOMOBRYUM JULACEUM

APHANORRHEGMA SERRATUM   Small ephemeral; round capsule opening by a slit, surrounded by the leaves

round capsules surrounded by leaves, opening on a line; capsule cell walls with thickened corners

A SMALL SUMMER AND FALL EPHEMERAL of moist open ground: cultivated fields, gardens, pastures, stream banks. Leaves oblong oval, with large clear cells, similar to those of *Funaria, Physcomitrium,* and *Physcomitrella.* Capsules round, almost stalkless, surrounded by the leaves, opening by a crosswise slit, the cell walls strongly thickened in their corners. Two rare species differ in their capsules. *Physcomitrium immersum* lacks the thickened corners. *Physcomitrella* has capsules that break irregularly rather than open on a line.

APHANORRHEGMA SERRATUM

ARCTOA   Small slender-leaved mosses of rock crevices in the alpine zone

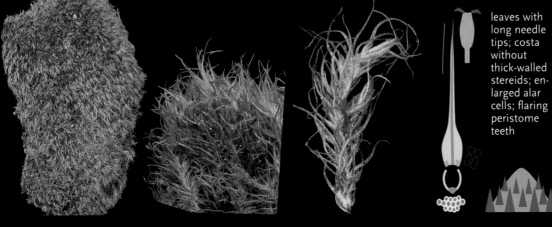

leaves with long needle tips; costa without thick-walled stereids; enlarged alar cells; flaring peristome teeth

SMALL MOSSES, FORMING TUFTS and small cushions between and under rocks in the alpine zone. In the field they much resemble *Dicranella heteromalla,* which may grow in the same habitat. The distinctions are microscopic: the alar cells of *Arctoa* are red and inflated, and the cells of the costa are all about the same size. We have two species, *fulvella* and *blyttii**, both rare and confined to the high mountains. Capsules are required to separate them.

*ARCTOA BLYTTII**

* Also called *Kiaeria blyttii*.

QUICK GUIDES TO HABITATS

QUICK GUIDES TO ACROCARPS

QUICK GUIDES TO PLEUROCARPS

QUICK GUIDES TO SPHAGNUM

ACROCARPS

PLEUROCARPS

SPHAGNUM

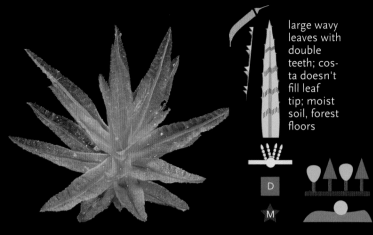

large wavy leaves with double teeth; costa doesn't fill leaf tip; moist soil, forest floors

COMMON MOSSES, forming large patches on moist, mineral soil, recognized by the lanceolate leaves, sharp teeth, long capsules, and lamellae on the costa. Abundant in moist forests and often forming bands along woodland trails and streams. Also in lawns, along roads, on tree bases and wet rocks, at the bases of ledges, and sometimes on open wet soil.

*ATRICHUM UNDULATUM GROUP**

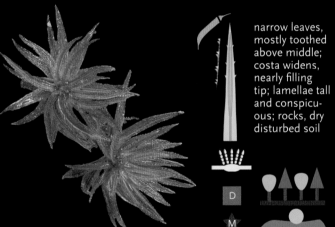

narrow leaves, mostly toothed above middle; costa widens, nearly filling tip; lamellae tall and conspicuous; rocks, dry disturbed soil

A SMALL SPECIES OF DISTURBED SOIL, boulders, and dry forest floors. Leaves shorter and narrower than the *undulatum* group, teeth mostly above the middle, often papillose on back; costa flaring upwards and tending to fill the narrow tip. The species is dioicous and male plants are common.

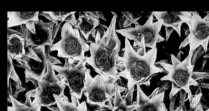

Male plants with antheridial cups

ATRICHUM ANGUSTATUM

broad leaves, usually with single teeth; ridges low, hard to see; stream banks, wet rocks, pond shores

A WETLAND SPECIES with broad leaves, and low inconspicuous ridges. Except for the ridges, it looks a lot like a *Mnium* or *Plagiomnium*. Locally common in sunny wet places: pond and stream shores, ditches, seeps. We find it regularly on old beaver lodges.

ATRICHUM CRISPUM

* The *Flora of North America* divides the group into three poorly demarcated species, *undulatum*, *altecristatum* and *crispulum*, separated by the leaf shape, height of the lamellae, and angle of the capsules. We find the distinctions hard to apply consistently

AULACOMNIUM Ⓕ Ⓜ Light-green toothed leaves; small round cells; often have gemmae

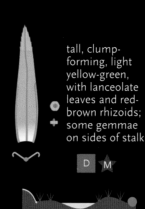

tall, clump-forming, light yellow-green, with lanceolate leaves and red-brown rhizoids; some gemmae on sides of stalk

DISTINCTIVE MOSSES: light-green or yellow-green toothed leaves and round, papillose cells. Two of our species have balls of gemmae on stalks. *Palustre*—big, yellow, and shaggy—is our largest and commonest species. It typically forms loose cushions among grasses and sedges in wetlands. It is a regular part of the bog flora, found at the tops of hummocks where it is too dry for *Sphagnum*.

AULACOMNIUM PALUSTRE

broad blunt-oval yellow-green rippled leaves; shoots somewhat flattened; big clumps on banks

A LARGE LEAFY MOSS, found on banks and tree bases in open woodland. Often associated with seepage. Recognized by the broad blunt leaves and somewhat flattened shoots. Young shoots are short and look like cabbages. A southern species, largely restricted to the oak zone.

AULACOMNIUM HETEROSTICHUM, with *Bartramia pomiformis*

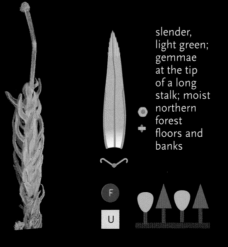

slender, light green; gemmae at the tip of a long stalk; moist northern forest floors and banks

A SMALL LIGHT-GREEN SPECIES, found on seepy banks and wet hollows in northern forest floors. Widespread in western North America. Uncommon but regular in the northern NFR, particularly near the Great Lakes and along the Atlantic coast. Recognized as an *Aulacomnium* by the toothed leaves and small round papillose cells; and as this species by the small size and ball of oval gemmae on a delicate stalk. *A. palustre* also has stalked balls of gemmae; it is much larger, and its gemmae are more triangular and occur on the sides of the stalk as well as the tip.

AULACOMNIUM ANDROGYNUM

QUICK GUIDES TO HABITATS

QUICK GUIDES TO ACROCARPS

QUICK GUIDES TO PLEUROCARPS

QUICK GUIDES TO SPHAGNUM

ACROCARPS

PLEUROCARPS

SPHAGNUM

BARBULA F E Small opaque mosses with parallel-sided leaves, multi-papillose cells, and twisted peristomes

leaf edges rolled under; tiny needle tips

BARBULA UNGUICULATA

SMALL YELLOWISH-OPAQUE MOSSES with papillose cells, parallel-sided leaves, a long tapering point on the capsule, and long twisted peristome teeth. They are calciphiles of open ground, found on rock, gravel, sand, concrete, or mortar, often in limy places. *Didymodon* species are similar in size and color and grow in similar places. Their leaves taper more to the tip, and they have shorter beaks on the capsules. Under the scope, *Didymodon* has less sharply defined papillae.

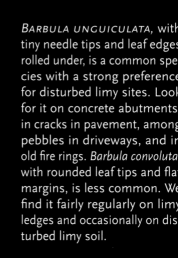

flat leaves with rounded tips

BARBULA CONVOLUTA

BARBULA UNGUICULATA, with tiny needle tips and leaf edges rolled under, is a common species with a strong preference for disturbed limy sites. Look for it on concrete abutments, in cracks in pavement, among pebbles in driveways, and in old fire rings. *Barbula convoluta*, with rounded leaf tips and flat margins, is less common. We find it fairly regularly on limy ledges and occasionally on disturbed limy soil.

BARTRAMIA POMIFORMIS C M Light-green slender-leaved moss with round capsules and red peristomes, in loose cushions on moist rocks

leaves with slender tips, double teeth, recurved edges, and short papillose cells; loose mounds on ledges and banks with seepage

A COMMON SPECIES OF MOIST SEEPY ROCKS, usually where the water (though not necessarily the soil) is calcareous. Pale white-green, with long leaf tips, forming open cushions and mossicles, also sometimes in mats. Small, round capsules are common. *Philonotis,* with shorter leaves, and *Plagiopus,* with red stems and shorter and darker leaves, have similar capsules and are related.

BLINDIA ACUTA (F) (U) Dark wiry moss on wet ledges or in the splash zone; red stem; slender erect leaves

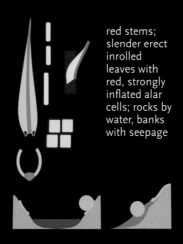

red stems; slender erect inrolled leaves with red, strongly inflated alar cells; rocks by water, banks with seepage

A SMALL, STRINGY MOSS found on wet rocks with seepage or in the spray zone of brooks. Found in both limy and acid sites. Recognizable in the field by the red stems and slender inrolled leaves. Often much darker than in the photos. Lab characters are the absence of teeth, relatively slender cells, and red thick-walled alar cells.

BLINDIA ACUTA

BRYOERYTHROPHYLLUM RECURVIROSTRUM (F) (U) Slender papillose leaves with recurved edges, old leaves reddish

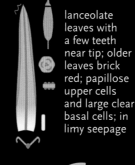

lanceolate leaves with a few teeth near tip; older leaves brick red; papillose upper cells and large clear basal cells; in limy seepage

A SMALL OPAQUE MOSS with oblong leaves that grows on wet limy rocks, often by rivers, and sometimes on concrete bridges. The brick-red old leaves are the best field character. *Didymodon, Barbula,* and *Trichostomum* look generally similar but do not have the brick-red color in the older leaves. When the color is weak, only the microscope can tell.

BRYOERYTHROPHYLLUM RECURVIROSTRUM, with *Encalypta procera,* the larger plants

BRYUM (M) Oval untoothed leaves with borders of long cells; capsule nodding, with a neck below the spore chamber

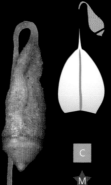

Pale white-green, in dense cushions on sand, rocks, and pavement. Leaves with small needle tips, upper cells empty

A LARGE GENUS OF SMALL MOSS-ES, usually clumped, often weedy, on mineral soil; leaves oval, short celled except for the borders; capsules nodding, often with a distinct neck. We show the common species and one rarity. *Argenteum* is tiny, common, and distinctive: tight budlike shoots, silver leaves, weedy in cracks in concrete and pavement. Compare *Anomobryum* and *Plagiobryum,* rare plants of cold limy ledges, also tight and silvery.

BRYUM ARGENTEUM

QUICK GUIDES TO HABITATS

QUICK GUIDES TO ACROCARPS

QUICK GUIDES TO PLEUROCARPS

QUICK GUIDES TO SPHAGNUM

ACROCARPS

PLEUROCARPS

SPHAGNUM

BRYUM CAPILLARE

budlike, in tight cushions; leaves broadly oval, the costa not reaching the tip; cells short and thin walled

A SMALL, VERY COMMON MOSS of intermittently moist ledges, banks, and shores. Field characters are the small budlike plants in dense cushions and broad oval leaves with small sharp tips. Technical characters are the short, bulging, thin-walled cells; costas often stopping below tip; and skinny brown gemmae in the leaf axils.

BRYUM LISAE GROUP

cells of medium length, not thin walled or bulging; costa excurrent as a slender tip

TWO WEEDY SPECIES, *lisae* and *caespiticium,* common in open habitats: rocks, mineral soil, disturbed places, even the high-tide edge of the coast. The costas that are excurrent into short needle tips and the cells over three times as long as wide identify the group. Some rare species (*amblydon, dichotomum*) are similar; you need capsules and a microscope to check for them.

plants tall, often pink; leaves short decurrent, with thin-walled cells; costa short excurrent; capsules with distinct necks

AN UNCOMMON SPECIES of shoreline seeps and rocks in streams. Plants tall rather than budlike; leaves decurrent, the costa reaching the tip or excurrent; cells short and thin walled; capsule with a long neck. May be a striking pink in spring.

BRYUM

BRYUM PSEUDOTRIQUETRUM

leaves slender, strongly bordered, long decurrent, often marked with red; cells thick walled

C M

A COMMON SPECIES OF WETLANDS AND SHORES, mostly in open sun, often mixed with other mosses. Also found in seepage on ledges and on shoreline rocks. Very common on hummocks in sedge meadows and fens, on shores with seepage, and on floating logs in ponds. Stems red; leaves strongly bordered and decurrent, the borders often red; cells short and relatively thick walled. The long-decurrent leaves are the best field character, the short cells a useful confirmation.

BUXBAUMIA F Tiny leaves from a persistent protonema; elongate, inclined capsules

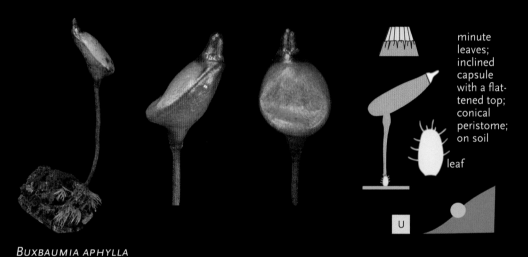

BUXBAUMIA APHYLLA

minute leaves; inclined capsule with a flattened top; conical peristome; on soil

leaf

U

UNUSUAL MOSSES with large, inclined, asymmetrical, stalked capsules. The plants, which are tiny, develop from a persistent protonema; the inner peristome of the capsule is fused into a conspicuous white pleated cone. *Diphyscium* has a similar peristome and often grows with *Buxbaumia*. Its capsules are stalkless and surrounded by long needle-pointed perichaetial leaves. *Aphylla*, with the top of the capsule flattened like an anvil, is a wide-ranging species, found throughout the NFR. It grows, regularly but uncommonly, on moist shaded soil banks, especially by brooks and along wood roads.

tiny oval leaves from a persistent protonema; elongate capsules, rounded in section, without flattened tops; on rotting logs

R

MINAKATAE, smaller and rarer, has rounded, tapering capsules without flattened tops and grows on rotting logs or organic soil over rocks. We find it, when we find it at all, as small groups of two or three capsules mixed with other mosses. It is known in eastern North America and eastern Asia. There are NFR records from MI, NY, MA, VT, ME, and NS.

BUXBAUMIA MINAKATAE, with *Dicranum flagellare* and squamules of *Cladonia*

QUICK GUIDES TO HABITATS

QUICK GUIDES TO ACROCARPS

QUICK GUIDES TO PLEUROCARPS

QUICK GUIDES TO SPHAGNUM

ACROCARPS

PLEUROCARPS

SPHAGNUM

CATOSCOPIUM NIGRITUM [R] [F] Small plants with lanceolate leaves in three rows; lower stems, setae, and capsules black

spaced lanceolate leaves in three rows; dark lower stems; tiny golf-club capsules

A RARE NORTHERN SPECIES of rich fens and limy seepage, found on cliffs and in fens. Transcontinental in Canada, coming south, sparingly, to the northern edge of the NFR. A small species, growing in clumps, sometimes hidden among other mosses. Leaves lanceolate, well spaced; lower stems black; capsules dark brown to black, like golf clubs. Fertile plants are distinctive. Resembles *Pohlia*, which is glossy and lacks three-ranked leaves, and *Meesia triquetra*, which has leaves that arch outwards.

CATOSCOPIUM NIGRITUM

CERATODON PURPUREUS [D] [M] [Ȇ] Lanceolate leaves with recurved edges and square cells; dark red-brown curved capsule

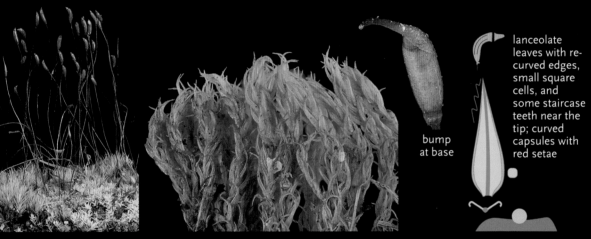

lanceolate leaves with recurved edges, small square cells, and some staircase teeth near the tip; curved capsules with red setae

bump at base

A COMMON SPECIES of open, sunny, disturbed mineral soil, fruiting abundantly; best recognized by the dark curved capsules with a small basal swelling and dark red setae. Sterile plants have light-green lanceolate leaves that twist up when dry, suggesting *Weissia* or one of its relatives. Microscopically, the recurved edges, small square cells, and low teeth near the tip are the best characters.

CERATODON PURPUREUS

CONOSTOMUM TETRAGONUM [F] [U] Cylindrical shoots with slender, light blue-green leaves in five vertical rows

  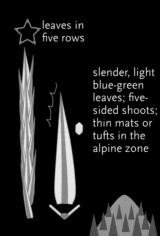

leaves in five rows

slender, light blue-green leaves; five-sided shoots; thin mats or tufts in the alpine zone

A SMALL, BRIGHT BLUE-GREEN MOSS of moist, protected places in the alpine zone, growing in tufts or mats, often amidst other mosses, often under overhangs or in cracks. The bright color and the five-rowed leaf arrangement, which makes the plants look like little stars when seen from above, are distinctive.

CONOSTOMUM TETRAGONUM

CYNODONTIUM TENELLUM Small slender-leaved moss found on cold sheltered rocks, especially at high elevations; cells papillose; upper margins thickened; alar cells not differentiated

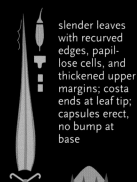

slender leaves with recurved edges, papillose cells, and thickened upper margins; costa ends at leaf tip; capsules erect, no bump at base

AN UNCOMMON NORTHERN SPECIES with slender long-tipped leaves that curl when dry, resembling a small *Dicranum*. Found on moist protected rock walls, in crevices, and under boulders, in eastern alpine zones and near the northern Great Lakes. Recognized microscopically by papillose cells, two-layered margins, and single teeth. The closely related *C. strumiferum* is similar but has an inclined capsule with a swelling at the base.

CYRTOMNIUM HYMENOPHYLLOIDES

CYRTOMNIUM HYMENOPHYLLOIDES Small blue-green plants with broadly oval untoothed leaves in flattened shoots

pale blue-green leaves with thin borders in arching flattened shoots

A SMALL PRETTY MOSS found in cracks and recesses in cold limy cliffs, usually where there is some seepage. Often associated with other cold-crevice species like *Anomobryum julaceum* and *Myurella julacea*. The pale color, flattened shoots, and bordered, broadly oval leaves are distinctive. Compare *Rhizomnium punctatum*, which has similar leaves but with heavier borders, spirally arranged shoots, and red on the older stems and leaves.

CYRTOMNIUM HYMENOPHYLLOIDES

DICHODONTIUM PELLUCIDUM A small linear-leaved papillose moss of wet rocks in and near streams

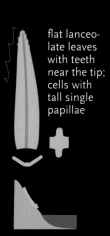

flat lanceolate leaves with teeth near the tip; cells with tall single papillae

A SMALL NONDESCRIPT MOSS, found erratically on wet rocks by streams, often with calcareous seepage. Leaves lanceolate, tapering from base to tip, not sharply folded. Similar to *Gymnostomum* in the field, but the leaves more tapering. Smaller than *Hymenostylium*, and *Amphidium*; very similar to *Didymodon*, but the leaves more sharply folded. The best lab characters are the tall papillae and the coarse teeth near the leaf tips.

DICHODONTIUM PELLUCIDUM

QUICK GUIDES TO HABITATS

QUICK GUIDES TO ACROCARPS

QUICK GUIDES TO PLEUROCARPS

QUICK GUIDES TO SPHAGNUM

ACROCARPS

PLEUROCARPS

SPHAGNUM

Leaves with costa excurrent into a long tip; alar cells not enlarged; seta yellow; capsules curved, contracted under the mouth when dry; common on soil banks, sometimes forming large patches

SMALL MOSSES, looking indeed like small *Di*cranums and also resembling *Cynodontium*, *Ditrichum,* and *Trematodon* vegetatively. Separated microscopically by the combination of slender leaves, undifferentiated alar cells, and forked peristome teeth. *Heteromalla,* with long slender leaf tips and capsules that are curved and contracted under the mouth when dry, is the only common species. *Schreberiana,* with hooked leaves that clasp at their bases, is a rare plant of limy wetlands. *Varia,* with recurved leaf edges and a red peristome and seta, is uncommon on open, episodically wet, limy soil.

DICRANELLA HETEROMALLA

leaves with clasping bases and long sharp-pointed tips that arch outward; limy seeps and wetlands

A DISTINCTIVE SPECIES, with arching leaves that clasp at their bases and long, pointed tips. It is northern and rare in our area. The few times we have seen it have been in limy fens and seeps; it is also reported from ditches and wet ledges. *Dicranella palustris,* reported from Maine and the Maritimes, has similar leaves with more rounded tips.

DICRANELLA SCHREBERIANA

slender leaves with parts of the edge recurved; red seta and peristome; open limy soil

A SMALL SPECIES, resembling *heteromalla,* that is locally common on moist open limy soil. It seems to be commonest in sites like river banks and borrow pits where there is some disturbance and the soil alternates between wet and dry. The best marks are the recurved leaf edges and the red peristome and seta.

DICRANELLA VARIA

* Four very rare species, *cerviculata, palustris, rufescens,* and *subulata,* are left out because we don't know enough to include them.

DICRANODONTIUM Slender leaves with broad midribs and inrolled edges, falling from older stems and leaving bare patches

slender tubular leaves with inflated alar cells and broad costas running into toothed needle tips; leaves fall from older branches

A WIDE-RANGING FOREST MOSS, common in the central Appalachians but rare in the NFR, with slender silky-looking leaves that can curve to one side. The leaves are oval at the base, inrolled above, and narrow to a long needle tip. The alar cells are inflated and clasp the stem. The costa is wide and not sharply distinguished from the leaf blade. The older leaves fall from the stems, leaving bare places. It is a forest generalist, growing on soil, stones, and logs, and resembles several other long-leaved forest species, particularly *Dicranella heteromalla*. The inflated alar cells, broad costa, and deciduous leaves are the best characters.

DICRANODONTIUM DENUDATUM

DICRANUM  Slender leaves, often with long tips and sharp teeth, with inflated reddish alar cells; peristome teeth forked

brood branches

concave leaves; erect capsules; brood branches; big or small mats on wood

COMMON MOSSES, often locally dominant, with lanceolate leaves, inflated alar cells, and forked peristome teeth. The first four species— subgenus *Orthodicranum*—have erect capsules and are relatively small and widespread. The remaining seven species have inclined capsules, are larger, and, except for *scoparium*, tend to be coastal or northern. *Flagellare*, with inrolled leaves without teeth and short deciduous brood branches with tiny leaves, is found on logs, tree bases, and shaded soil. It makes large patches and is one of our commonest woodland mosses. The brood branches are the best character; usually you can find at least a few.

DICRANUM FLAGELLARE

Leaves two-cells thick in upper parts, papillose below, with relatively broad costas; common on rocks and ledges, occasional on wood

ONE OF OUR COMMONEST DRY-ROCK MOSSES, making large dark-green mats on boulders and ledges. Similar to *flagellare* but usually curlier; very similar to *scoparium* and often mixed with it. The most reliable characters are the erect capsules, the broad costas, and the bistratose (two-layered) upper parts of the leaves.

DICRANUM FULVUM

QUICK GUIDES TO HABITATS

QUICK GUIDES TO ACROCARPS

QUICK GUIDES TO PLEUROCARPS

QUICK GUIDES TO SPHAGNUM

ACROCARPS

PLEUROCARPS

SPHAGNUM

mostly < 1 cm high; tightly curled when dry

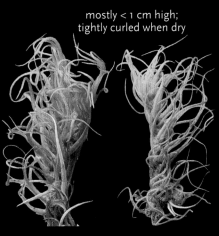

DICRANUM MONTANUM

small plants in dense dark mats or cushions; leaf cells in upper parts of leaf papillose below

SMALL DARK PLANTS, tightly curled when dry, the upper cells papillose below. Very common in small or large cushions or mats on logs, bark, and boulders. Presumptively identifiable by the small size and slender curly leaves; confirmation requires seeing the papillae on the leaf cells.

DICRANUM VIRIDE

medium-sized plants, in small cushions or scattered among other mosses; leaves straight, channeled, untoothed, with broken tips

LIGHT-GREEN PLANTS, not curly when dry, with broken leaf tips. Common but rarely abundant, usually in small tufts, on tree trunks and bases, on wet logs in ponds, and in shaded wetlands. The straight, stiff, widely spreading leaves with the tips broken off are distinctive.

DICRANUM POLYSETUM

straight shiny light-green leaves with ripples; multiple setae

THE REMAINING DICRANUMS are medium-sized to large species with inclined capsules. *Polysetum* is a large woodland species, often locally dominant. It is very common in the boreal forest where, with *Polytrichum commune, Hylocomium splendens, Ptilium crista-castrensis,* and *Hypnum imponens,* it makes a continuous ground layer. The straight shiny ripply leaves and multiple setae are distinctive. *D. scoparium,* which often grows with it, lacks ripples and often has curved leaves. *D. ontariense* has weaker ripples and is strongly curled when dry.

DICRANUM

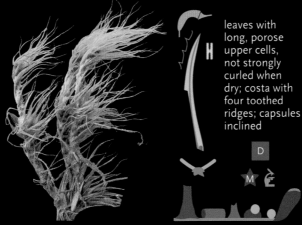

leaves with long, porose upper cells, not strongly curled when dry; costa with four toothed ridges; capsules inclined

COMMON AND CONSPICUOUS, making cushions on rocks and on the forest floor. Presumptively recognizable in the field by the long leaves curved in one direction that do not curl when dry, sharp teeth, and single setae. *Ontariense*, *fuscescens*, and some northern and alpine rarities are similar and best separated by the microscope. Good characters for *scoparium* are the elongate upper cells and the strongly ridged and toothed costa. Separation sometimes fails: all the critical characters vary, and puzzling specimens are common.

DICRANUM SCOPARIUM

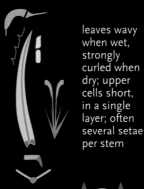

leaves wavy when wet, strongly curled when dry; upper cells short, in a single layer; often several setae per stem

AN EASTERN-NORTH AMERICAN SPECIES of forest floors, commonest in the conifer and mixed woods of the Maritime Provinces and northern Great Lakes. Resembles *scoparium* and *fuscescens* in the field. The best field characters are the multiple setae and the leaves that are slightly wavy when wet and strongly curled when dry. Microscopic confirmation is prudent: the short upper cells separate it from *scoparium*, and the unthickened margins from *fuscescens*.

DICRANUM ONTARIENSE

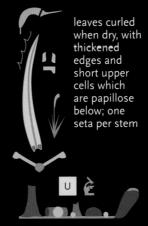

leaves curled when dry, with thickened edges and short upper cells which are papillose below; one seta per stem

A HARD-TO-RECOGNIZE NORTHERN SPECIES of stumps, logs, and the forest floor, with single setae and leaves that curl when dry. Locally common near the forest floor in northern mixed forests, especially in the mountains and near the Great Lakes and the coast. Uncommon southwards and at low elevations. Large plants resemble *Dicranum scoparium*; small ones are much like *fulvum* or *flagellare*. The safest characters are microscopic: thickened upper leaf edges and short upper cells that are papillose below.

DICRANUM FUSCESCENS

QUICK GUIDES TO HABITATS

QUICK GUIDES TO ACROCARPS

QUICK GUIDES TO PLEUROCARPS

QUICK GUIDES TO SPHAGNUM

ACROCARPS

PLEUROCARPS

SPHAGNUM

leaves oval at the base, broader than most other *Dicranums,* concave above; growth in segments with interruptions; upper cells small, thick walled, papillose below

F U

A MEDIUM-SIZED SPECIES OF DRY SOIL, most often seen among lichens in sandy or rocky barrens. It is wide ranging in northeastern North America and seems most common along the Atlantic coast. The oval bases of the leaves and interrupted or tufted appearance of the shoots are the best field characters; the small thick-walled angular upper cells that are papillose below are a good confirmation. *Dicranum condensatum,* a similar species with keeled rather than concave leaves, is found along the southern edge of the NFR.

DICRANUM SPURIUM

restricted to bog hummocks; growth in segments; leaves wavy, keeled, erect, often blunt at extreme tip, often twisted together at stem tips; costa ending below tip

F U

A WIDE-RANGING NORTHERN SPECIES that grows with *Sphagnum* in hummocks in bogs. Typically large, light yellow-green, with undulate leaves that, when dry, are straight and either erect at the stem tips or twisted into a point at the stem tips. The upper leaf cells are short, the costa ends below the tip, and the leaves are separated from those of the previous year. This segmented growth pattern occurs in other Dicranums, particularly *spurium* and *bonjeanii. Spurium* is broader-leaved and usually in dry habitats. *Bonjeanii* looks a lot like *undulatum* and can grow with it in bogs. It is distinguished under the microscope by the longer upper cells.

DICRANUM UNDULATUM

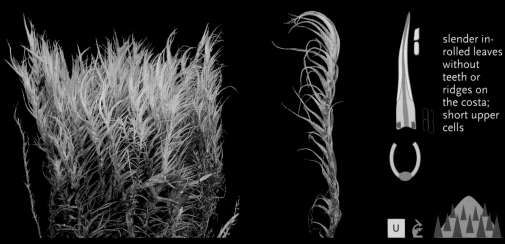

slender inrolled leaves without teeth or ridges on the costa; short upper cells

U ≩

A FAR-NORTHERN SPECIES reaching the Arctic Ocean in Alaska and Nunavut, rare on alpine summits in the NFR. Plants delicate and slender, often shiny, in dense tufts, with the upper edges of the leaves inrolled, making a channeled or tubular tip. Leaves untoothed, with short smooth upper cells and no ridges on the costa. Very rare in the NFR. We have alpine records from Katahdin, the Chic-Chocs, the White Mountains, and the High Peaks, plus a few low elevation records from Nova Scotia and Minnesota.

DICRANUM ELONGATUM

DICRANUM

leaves straight, in-rolled, weakly toothed, often undulate; costa with two ridges; upper cells porose, somewhat wavy

A PROBLEMATIC SPECIES, close to *scoparium* and, according to Ireland's treatment in the *Flora of North America*, often confused with it. As used here, an uncommon species, often on hummocks in bogs and fens, resembling *scoparium* but with straighter leaves that are only weakly toothed and have two rather than four ridges on upper the costa. The leaves can be ripply and resemble *Dicranum undulatum*. The upper cells of *undulatum* are often shorter, and its leaves are twisted into a point at the stem tips when dry.

DICRANUM BONJEANII

leaves > 9 mm long; several setae per stem; costa with two rows of guide cells in cross section

A WIDE-RANGING CIRCUMBOREAL SPECIES of northern coasts, found mostly in the arctic and subarctic. In the NFR it is found in moist conifer woods and stream ravines near the Atlantic coast in Maine, the Maritimes and, rarely, near the Great Lakes. In the field it resembles a large *Dicranum scoparium* with leaves over 9 mm long and, often, several setae per stem. Microscopically it is separated from *scoparium* by leaf cross-sections: *majus* has two rows of guide cells, *scoparium* has one.

DICRANUM MAJUS, with the liverwort *Bazzania trilobata*

DIDYMODON U ℮ Small pale-green mosses of moist limy soil, with tapering leaves and low hard-to-see papillae

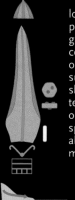

low obscure papillae; elongate cells of costa exposed on upper surface; leaves sharply folded, tending to lie on their sides, springing open abruptly when moistened

SMALL DIFFICULT MOSSES of limy soil and ledges, common but not abundant in the proper habitat. In the field, they grow in small, loose tufts or groves and have leaves that tend to be broadest near the base and taper evenly to the tip. The leaf shape helps separate *Didymodon* from related genera; *Barbula* has nearly parallel-sided leaves; *Gymnostomum* and *Hymenostylium* have somewhat narrower leaf bases than *Didymodon*.

*DIDYMODON FERRUGINEUS**

* Formerly *Barbula reflexa*.

QUICK GUIDES TO HABITATS

QUICK GUIDES TO ACROCARPS

QUICK GUIDES TO PLEUROCARPS

QUICK GUIDES TO SPHAGNUM

ACROCARPS

PLEUROCARPS

SPHAGNUM

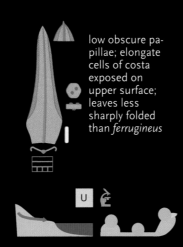

low obscure papillae; elongate cells of costa exposed on upper surface; leaves less sharply folded than *ferrugineus*

THE SPECIES OF DIDYMODON require a microscope and experience. As a group, they are recognized by the small rounded cells, recurved edges, and by having at least a few low papillae on the upper cells. *Ferrugineus* and *fallax* have the elongate cells of the costa visible on the upper surface of the leaf (▦). *Rigidulus* has the costa covered by short cells on the upper surface (▦). *Ferrugineus* is distinguished by strongly reflexed leaves that are sharply folded for much of their length and often have taller, more distinct papillae. *Fallax* has spreading rather than reflexed leaves that are less strongly folded and often have lower, less distinct papillae. The two are close.

*DIDYMODON FALLAX**

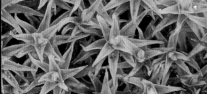

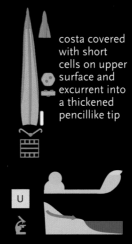

costa covered with short cells on upper surface and excurrent into a thickened pencillike tip

A WIDESPREAD SPECIES, most commonly found on limy rock by streams. Recognized under the microscope by the short cells covering the costa on the upper side of the leaf, the excurrent costas, and thickened tips. Our other species of *Didymodon* have the long cells of the costa exposed above and costas that end in the leaf tip.

DIDYMODON RIGIDULUS

DIPHYSCIUM FOLIOSUM [C] [M] A small common moss with a big asymmetrical stalkless capsule

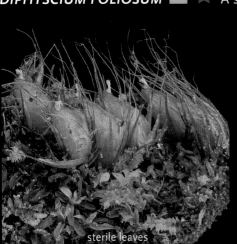

sterile leaves

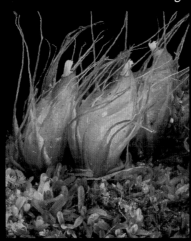

straplike oblong opaque leaves; fat lopsided capsules surrounded by needle-tipped perichaetial leaves; moist soil banks in the shade

A COMMON SPECIES of shaded mineral soil, often on banks in woods. Sterile shoots are stalkless, with straplike leaves and large papillae that obscure the cells. Fertile plants have long perichaetial leaves with needle tips and a bulgy penguin-shaped capsule.

DIPHYSCIUM FOLIOSUM

* *Didymodon fallax* was formerly *Barbula fallax*; *Didymodon tophaceus* (not shown), a southern species with blunt leaves that grows in seepage and is often lime-encrusted, occurs rarely in the NFR.

DISTICHIUM Tiny mosses of cold limy crevices in the mountains; slender leaves with long-clasping white-green bases

needle-pointed leaves with clasping bases, in two rows; capsules straight, erect

TWO DELICATE MOSSES with slender leaves arranged in two rows, like corn. Southwards found in small amounts on limy soil in crevices, often by rivers; northwards found in dense cushions in limy seeps and barrens. Our wide-spread species, *capillaceum,* has erect capsules; *inclinatum,* rare and far northern, has inclined ones.

DISTICHIUM CAPILLACEUM

DITRICHUM C M Weedy slender-leaved mosses of mineral soil, with long setae, long erect capsules, and papillose peristome teeth that are divided nearly to the base

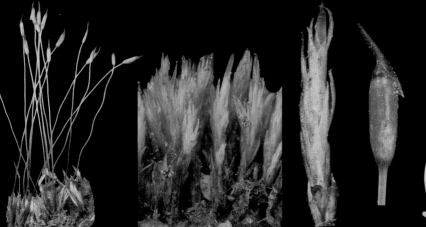

main leaves with long tips; side branches with reduced leaves with short tips; reddish seta; peristome teeth uniformly papillose

DITRICHUMS are small narrow-leaved mosses of open disturbed mineral soil, forming tufts or mats and sometimes covering large areas. Fruiting plants are recognized by their long setae and capsules, and, under the scope, by slender papillose peristome teeth that are divided nearly to their bases. Sterile plants look much like several other genera of disturbed soil, particularly *Dicranella, Bruchia, Pleuridium,* and *Trematodon.* They can be guessed at but not accurately determined.

DITRICHUM LINEARE

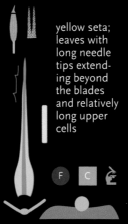

yellow seta; leaves with long needle tips extending beyond the blades and relatively long upper cells

DITRICHUM LINEARE AND *PALLIDUM* are the common NFR species. *Lineare* has erect capsules and reddish setae, and often has short vegetative branches with small blunt leaves. *Pallidum* has long slender needle-tipped leaves, bright yellow setae, and capsules with short beaks. The leaves have longer tips than those of *lineare* and *pusillum. Dicranella heteromalla,* which grows in similar habitats, has leaves and setae like *pallidum* but curved capsules with long beaks.

*DITRICHUM PALLIDUM**

* *Ditrichum pusillum,* not shown, is a similar species that is separated from *pallidum* by the red seta and from *lineare* by having peristome teeth that are more completely divided and marked with oblique lines.

QUICK GUIDES TO HABITATS

QUICK GUIDES TO ACROCARPS

QUICK GUIDES TO PLEUROCARPS

QUICK GUIDES TO SPHAGNUM

ACROCARPS

PLEUROCARPS

SPHAGNUM

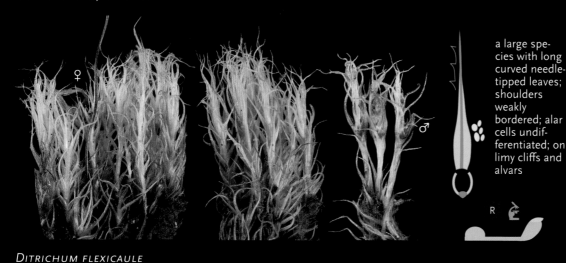

a large species with long curved needle-tipped leaves; shoulders weakly bordered; alar cells undifferentiated; on limy cliffs and alvars

FLEXICAULE is a northern species, large for a *Ditrichum,* typically found in thin soil over cold calcareous ledges or in crevices and shallow pools in limestone barrens. In the field it looks like an oversized *Dicranella.* Its best marks are the furry lower stems, the slender concave leaves with long needle tips, the undifferentiated alar cells, and the small obliquely oriented cells that form a weak border at the shoulders.

DITRICHUM FLEXICAULE

DRUMMONDIA PROREPENS F A tree-bark moss, resembling *Orthotrichum,* that forms colonies by creeping stems; leaf edges flat

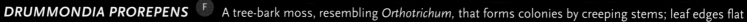

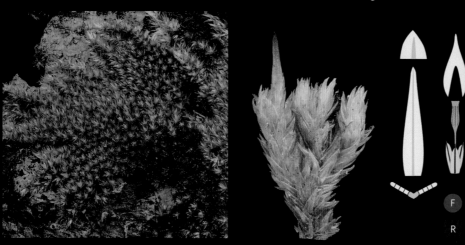

small species, resembling an *Orthotrichum,* forming patches by creeping; leaves lanceolate with flat edges; calyptra hairless

AN APPALACHIAN SPECIES of tree bark, rare with us but found occasionally in the southern NFR and along the New England coast. Basically a smooth-celled *Orthotrichum* that creeps: lanceolate leaves with flat margins; erect furrowed capsules; and creeping stems that form patches or traceries. We have found it on large maples along roads. There are many more maples without *Drummondia* than with, and, in fact, very few recent records from the NFR.

DRUMMONDIA PROREPENS

ENCALYPTA C M Short-stalked mosses with broad strap-shaped yellow-green papillose leaves, commonly on limy rocks or soil

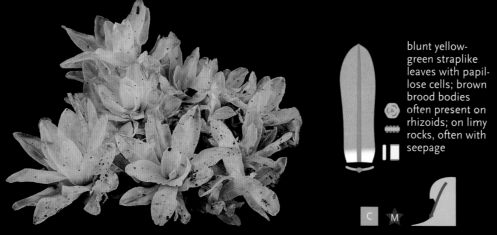

blunt yellow-green straplike leaves with papillose cells; brown brood bodies often present on rhizoids; on limy rocks, often with seepage

SMALL TO MEDIUM-SIZED MOSSES with relatively short stalks and wide opaque yellow-green leaves with multipapillose cells. The lower leaf cells have strongly thickened, red cross-walls. All are associated with ledges and limy seepage. *Procera* is our common species, usually sterile, with brown brood bodies and blunt vegetative leaves. Fruiting plants, which are rare, have short needle tips on the leaves below the setae; stems have central strands. It grows in thin soil on limy rocks, often in seepage cracks, and often with *Tortella tortuosa, Mnium marginatum, Gymnostomum aeruginosum,* or *Syntrichia ruralis.*

ENCALYPTA PROCERA

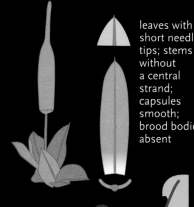

leaves with short needle tips; stems without a central strand; capsules smooth; brood bodies absent

ENCALYPTA CILIATA looks like *procera* and grows in the same sort of places but is smaller, more northern, more often fertile, and much less common. When fertile, the elongate calyptra makes it an *Encalypta* and the smooth capsule makes it *ciliata*. Vegetative plants are identified by the presence of short needle tips on at least some of the leaves and the absence of a central strand in the stem.

ENCALYPTA CILIATA

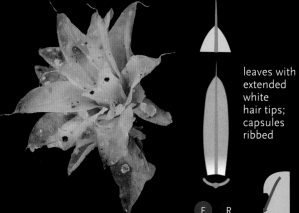

leaves with extended white hair tips; capsules ribbed

A SMALL, FAR-NORTHERN SPECIES, quite rare with us, recognized by the hair tips of the leaves and the straight-ribbed capsules. Note, for safety, that our common *Encalypta procera*, though rarely fertile, has spiralling ribs on its capsules and needle tips on the leaves just below the seta. The few times we have seen *rhaptocarpa*, it has been on limy ledges—Grenville-age marble—in the splash zone of rivers and lakes. Farther north, beyond the NFR, it is found in seeps and limy barrens.

ENCALYPTA RHAPTOCARPA

EPHEMERUM Tiny mosses of open limy soil with slender toothy leaves, round capsules, and persistent bright emerald-green protonema

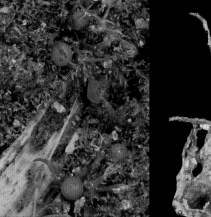

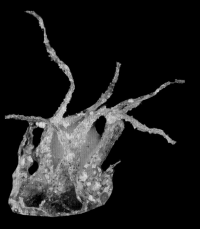

plants minute; leaves sharp toothed, from a persistent green protonema; capsules round

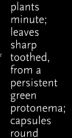

TINY UNCOMMON PRETTY MOSSES of open moist fertile soil, found in cultivated fields, wet pastures, and dried-up ponds. The plants are tiny and stalkless, with slender leaves with distinctive needle teeth. They develop from a velvety bright-green protonema that can be quite striking in the field. *Ephemerum* is southern and uncommon in the NFR. We have four species—*cohaerens, crassinervium, serratum, and spinulosum*—only distinguishable with the microscope.

EPHEMERUM CRASSINERVIUM

QUICK GUIDES TO HABITATS

QUICK GUIDES TO ACROCARPS

QUICK GUIDES TO PLEUROCARPS

QUICK GUIDES TO SPHAGNUM

ACROCARPS

PLEUROCARPS

SPHAGNUM

FISSIDENS C M Flattened mosses with the leaves in two rows; leaves have pockets along the upper edge

large and fernlike; leaves with pale borders; teeth jagged and irregular; cells flat, in a single layer

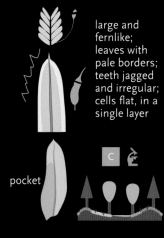

pocket

FISSIDENS ADIANTHOIDES

A LARGE WARM-TEMPERATE AND TROPICAL GENUS, varying in size and leaf shape but always with flattened shoots with the leaves in two rows and a pocket on the upper edge of the leaf that usually clasps the leaf above. Found most commonly near the ground, on rocks, soil, and logs. Often in wetlands, along streams and shores, and in wet woods. Easily recognized to genus in the field; the species require a microscope. *F. adianthoides* is our largest species, growing in big clumps in wet woods and forested and open wetlands. Its marks are the jagged, irregular teeth; a faint border of paler cells; and flat cells in a single layer. *Dubius* is similar but has smaller cells that sometimes overlap.

medium sized; leaves with pale borders; teeth jagged and irregular; cells small, bulging, sometimes overlapping or two-layered

FISSIDENS DUBIUS

DUBIUS is a medium-sized species, very common on moist soil and rocks, also on tree bases and occasionally on dry ledges or boulders. In good habitats like rocks in the splash zone of streams, it can form large patches. The critical characters are the jagged teeth and the pale borders of the leaves, which can often be seen with a 20× lens; and the small bulging cells that occasionally overlap each other, which require a microscope.

teeth low and even; costa ends a few cells below leaf tip; seta from tip of shoot

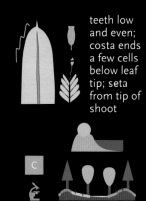

A COMMON SPECIES, medium sized, similar to *dubius* and found in the same sort of places. Distinguished, like *taxifolius* and *bushii*, by its lower, more regular teeth. Separated from them by the capsules from the tips of shoots and the costa that stops several cells below the tip.

QUICK GUIDES TO HABITATS

QUICK GUIDES TO ACROCARPS

QUICK GUIDES TO PLEUROCARPS

QUICK GUIDES TO SPHAGNUM

ACROCARPS

PLEUROCARPS

SPHAGNUM

FISSIDENS

small; setae from base of fronds; low even teeth; costa ends at tip; cells of the pocket with several papillae

FISSIDENS BUSHII

Bushii AND *TAXIFOLIUS* are small species, under 1 cm high, most often seen on bare soil in woods or in brushy fields or near the ground on rocks, and tree bases. They are fairly common but need to be looked for. Both have even teeth, papillae on the cells of the pocket, and capsules from the base of the stem. *Bushii* has a weak costa and multipapillose cells on the edge of the sheathing part of the leaf.

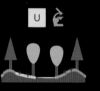

small; setae from base of fronds; low even teeth; costa sticks out at tip; cells of the pocket with a single papilla

FISSIDENS TAXIFOLIUS

TAXIFOLIUS is similar to *bushii* and grows in the same sort of habitats; we are not able to tell them apart in the field or predict which we will see where. Its best marks are microscopic: the costa which sticks out at the leaf tip and the single papillae on the cells of the pocket.

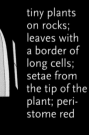

tiny plants on rocks; leaves with a border of long cells; setae from the tip of the plant; peristome red

FISSIDENS BRYOIDES

A TINY MOSS, commonly under 5 mm high and barely visible unless fruiting, found most commonly on rocks in streams but also on rocks in moist woods. The small erect capsules and red peristomes are good marks. Sometimes all you see are the peristomes. Fruiting plants can be identified by the terminal setae; in our other small species the setae come from the base of the plant. Sterile plants are identified microscopically by the border of elongate cells around the leaves.

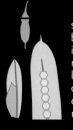

small, often brown green, in dense mats on limestone; costa covered with round cells

A SMALL SPECIES of intermittently wet limestone ledges, growing in flat, dark mats and sometimes covering large areas; southern and rare in the northern forest. Presumptively identifiable in the field by the dense dark mats; confirmation requires seeing the small round cells that cover the costa.

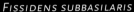
FISSIDENS SUBBASILARIS

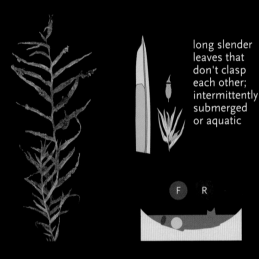

long slender leaves that don't clasp each other; intermittently submerged or aquatic

LARGE AQUATIC SPECIES, usually growing on wood in ponds and slow streams, submerged or exposed at low water. Leaves long and slender, widely separated, the pouches not clasping the leaf above. A southern species, restricted to the oak zone and rare with us. *Dichelyma,* which can grow with it and also has slender leaves, is three ranked and the leaves don't have pouches.

FISSIDENS FONTANUS

FUNARIA HYGROMETRICA D M Small short-lived moss of open disturbed soil; plants budlike; capsules curved and nodding

A COMMON EPHEMERAL, found throughout North America. Small plants, germinating and fruiting rapidly on disturbed open soil, often forming large patches after fires. Recognized by the arching setae and distinctive asymmetrical capsules. The pale oblong-oval leaves have large thin-walled cells and resemble those of *Physcomitrium, Aphanorrhegma,* and *Physcomitrella.*

FUNARIA HYGROMETRICA

GRIMMIA C M 𝔢

Dark mosses with slender leaves, forming cushions on rock; lanceolate or linear leaves, often with white needle or hair tips; short smooth thick-walled cells; perichaetial leaves not enlarged

QUICK GUIDES TO HABITATS

QUICK GUIDES TO ACROCARPS

QUICK GUIDES TO PLEUROCARPS

QUICK GUIDES TO SPHAGNUM

ACROCARPS

PLEUROCARPS

SPHAGNUM

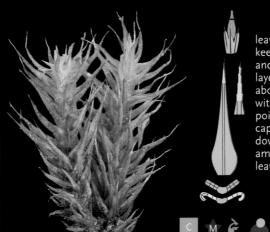

leaves keeled and two-layered above, with long points; capsules down among leaves

THE GRIMMIAS are small mosses in dense cushions, always on rock, often in the open, often with white needle points. There are many species in northern and western North America but only a few in the NFR.* Field marks are the dark color, the dense clumps, the lance-shaped leaves which often have white points, and the short oval capsules. *Racomitrium* is looser and branchier and has distinctive wavy-edged cells. *Schistidium*, which was formerly included in *Grimmia*, has enlarged perichaetial leaves at the base of the seta and at most short white tips.

GRIMMIA PILIFERA

leaves keeled, with long points; capsules above leaves; cells on lower margins elongate

PILIFERA, above, is a common Appalachian species of exposed acid rocks. The combination of long hair-points and capsules surrounded by the leaves is distinctive. It is an oak-zone species and largely restricted to the southern parts of the NFR.

Donniana, left, the common *Grimmia* of the alpine zone, has keeled leaves, long needle tips, and exserted capsules. Similar plants occurring at lower elevations in Maine, the Adirondacks, and the Great Lakes may also be *donniana* or the closely related *incurva*. The Adirondack ones look like *donniana* to us.

GRIMMIA DONNIANA

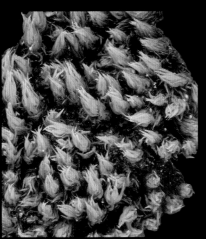

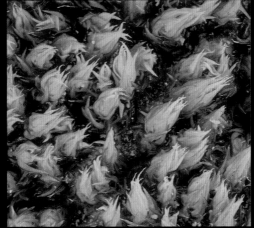

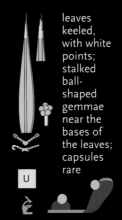

leaves keeled, with white points; stalked ball-shaped gemmae near the bases of the leaves; capsules rare

U

A TINY *GRIMMIA* found on boulders and stone walls; leaves keeled, margins recurved on one side, stalked ball-shaped gemmae on the bases of the leaves. Locally common in Maine and the western Great Lakes; rare otherwise. In the field it is a very small *Grimmia*, only a few millimeters high, on dry rocks. The gemmae, best seen with a microscope, are the diagnostic feature.

GRIMMIA MUEHLENBECKII

* Rare species, not treated here, include *longirostris*, like *pilifera* but with capsules above the leaves; *incurva*, like *donniana* but with a recurved edge; and *hartmannii*, like *muehlenbeckii* but with the gemmae at the leaf tips.

concave
leaves with
conspicuous
white tips,
several cells
thick near the
tip; capsules
on curved
setae

AN APPALACHIAN SPECIES, uncommon in the NFR, usually found by water, on both acid and limy rocks. Small plants in cushions, with curved setae and conspicuous white leaf tips. The curved setae and nodding capsules, when present, are unique among our species. Sterile plants are recognized by the concave leaves that have oval bases and are several cells thick above.

GRIMMIA OLNEYI

blunt,
concave
leaves with
thickened
canoe-tips;
on rocks by
water

A SMALL NORTHERN SPECIES found on wet cliffs and rocks in the spray zone of lakes and rivers. Widely distributed along the shores of the northern Great Lakes; much rarer in the eastern NFR. Recognized by the concave leaves with hooded tips that are several cells thick. *Schistidium agassizii* looks similar and is said to grow in the same habitat; its tips are not thickened, and its capsules are shorter. We have looked for it but never found it.

GRIMMIA UNICOLOR

GYMNOSTOMUM AERUGINOSUM

Small papillose mosses with flat leaf edges and a costa ending below the tip; leaves lanceolate; costa covered with short cells above

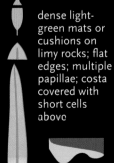

dense light-
green mats or
cushions on
limy rocks; flat
edges; multiple
papillae; costa
covered with
short cells
above

A SMALL LIGHT-GREEN MOSS that grows in protected places on limy rocks, often in seepage, sometimes making thick mats, sometimes pendent in mossicles. Similar to *Amphidium*, *Hymenostylium*, *Barbula*, and *Rhabdoweisia*. The diagnostic features are the flat edges, multiple papillae, and costa covered by short cells above and ending below the tip.

GYMNOSTOMUM AERUGINOSUM, with *Encalypta procera*

HEDWIGIA CILIATA  D M

Large moss in loose patches on ledges and boulders; stringy when dry, bristly when wet; concave leaves with papillose cells, white needle tips, and no costa

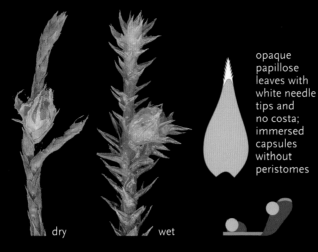

opaque papillose leaves with white needle tips and no costa; immersed capsules without peristomes

dry / wet

A COMMON SPECIES of boulders and dry ledge crests. Gray green, opaque, and skinny when dry, with the white needle tips conspicuous. Yellow green, translucent, and wide spreading when wet, with the tips harder to see. Often with *Schistidium*, *Grimmia*, *Dicranum scoparium*, *Dicranum fulvum*, and *Polytrichum piliferum*. Dwarfed northern forms from exposed rocks can be nearly black and look like *Racomitrium*.

HEDWIGIA CILIATA

HYMENOSTYLIUM RECURVIROSTRUM 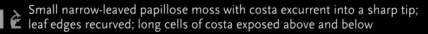 C

Small narrow-leaved papillose moss with costa excurrent into a sharp tip; leaf edges recurved; long cells of costa exposed above and below

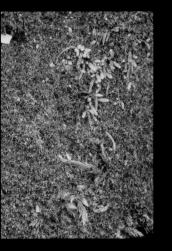 

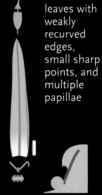

narrow leaves with weakly recurved edges, small sharp points, and multiple papillae

A SMALL MOSS with linear lanceolate, keeled, papillose leaves, growing in dense masses on cold wet limy cliffs. Often in recesses where there is seepage; can hang from the roofs of crevices and form mossicles. Similar to *Gymnostomum* and *Amphidium* and can grow with them; we don't try to distinguish them in the field.

HYMENOSTYLIUM RECURVIROSTRUM

HYOPHILA INVOLUTA  F R

Tiny moss of rocks in and near streams; oval leaves, narrowed to base, that roll into tubes as they dry

oval papillose leaves, rounded at their tips, that roll inwards from their edges, like a scroll, when they dry

A SMALL MOSS with broadly oval leaves, found on limy rocks in streams. Rare with us, common southward. Young plants are silvery; old plants may be blackish. The leaves are untoothed, have blunt tips, and roll up into tubes as they dry. No other stream mosses look anything like it.

HYOPHILA INVOLUTA

QUICK GUIDES TO HABITATS

QUICK GUIDES TO ACROCARPS

QUICK GUIDES TO PLEUROCARPS

QUICK GUIDES TO SPHAGNUM

ACROCARPS

PLEUROCARPS

SPHAGNUM

LEPTOBRYUM PYRIFORME Small moss with needle-tipped leaves and nodding capsules, forming clumps on disturbed soil

needle-tipped leaves with long cells and clasping bases; nodding capsules abruptly narrowed to long slender necks

A SMALL LIGHT-GREEN MOSS of disturbed soil, found throughout North America. Common as a greenhouse weed in the NFR, widespread but uncommon on open moist mineral soil. The nodding capsules with long necks are distinctive; sterile plants resemble *Dicranella* but have longer cells.

LEPTOBRYUM PYRIFORME

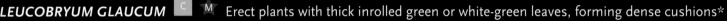

LEUCOBRYUM GLAUCUM C M Erect plants with thick inrolled green or white-green leaves, forming dense cushions*

pale inrolled leaves, several cells thick, in dense cushions

ONE OF THE COMMONEST MOSSES of the forest floor in both deciduous and boreal woods. Also on tree roots, well-decayed logs, and hummocks in swamps. The leaves are essentially all costa: several cells thick, with small green cells surrounded by larger water-holding ones. Forms small mounds in dry woods, larger patches and hummocks in wet ones.

LEUCOBRYUM GLAUCUM

MEESIA F R Curved capsules with long necks; square cells; leaves often in rows, arching outwards from a clasping base

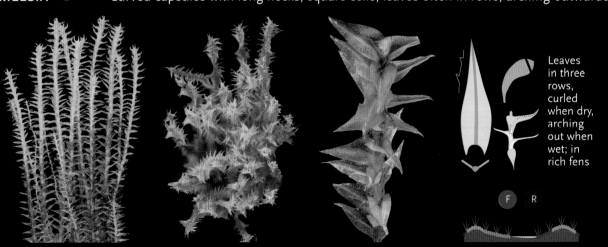

Leaves in three rows, curled when dry, arching out when wet; in rich fens

AN UNCOMMON MOSS, found in small amounts on hummocks or around the bases of shrubs in rich fens, distinguished by the arching leaves in three rows. It is a subarctic species, rare this far south. Two other Meesias, *longiseta* (not shown) and *uliginosa*, both northern and restricted to fens, are even rarer.

MEESIA TRIQUETRA

* *Leucobryum albidum*, not shown, is either a small form of *glaucum* or a sister species to it. Paul Redfearn says in the *Flora of North America* that it is southern, makes small clumps, has reflexed leaf tips, and intergrades with *glaucum*. We think of *glaucum* as a variable species and don't use *albidum*.

MEESIA

straplike leaves with recurved edges, broad costas, blunt tips, and rectangular cells; curved capsules with long necks

A SMALL RARE MOSS of peaty soil with calcareous seepage, found on cliffs, shores, and in fens. The narrow leaves with recurved edges, broad costas, and narrow rounded tips are unique. The species is northern and at its southern limits here. It is scattered in the northern Great Lakes and almost absent from the eastern NFR. Our photos came from the Keweenaw Peninsula.

MEESIA ULIGINOSA

MNIUM C M ξ Elliptical leaves, mostly bordered, with double teeth that may be hard to see; capsules angled down

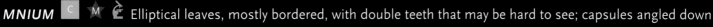

Narrow-elliptic decurrent leaves; costa toothed below, stops short of tip

COMMON AND CONSPICUOUS MOSS-ES of moist forest floors and ledges, with elliptical leaves with paired teeth and (except *stellare*) a border of linear cells. The teeth of the larger species can be seen with a 20× lens; those of the small species need a microscope.

Hornum, our commonest species, is recognized by the large narrow leaves and found in all kinds of wet shaded places: hollows, tree bases, soil banks, and, especially, in conspicuous bands along the banks of woodland streams.

MNIUM HORNUM

strong teeth; costa not toothed; multiple setae; peristome dark red; in mixed and conifer woods

A COMMON SPECIES of moist conifer forests, especially on fertile soils. Wide ranging in northern North America and found throughout the NFR. Oval leaves with conspicuous teeth, smooth costas, and multiple setae. The capsules appear in late summer, after our other Mniums, and are light tan with dark red peristome teeth. This and *M. hornum* are the two medium-sized Mniums regularly seen on forest floors. *Hornum* has longer and narrower leaves, a toothed costa that ends below the tip, single setae, and yellow peristomes.

MNIUM SPINULOSUM

QUICK GUIDES TO HABITATS

QUICK GUIDES TO ACROCARPS

QUICK GUIDES TO PLEUROCARPS

QUICK GUIDES TO SPHAGNUM

ACROCARPS

PLEUROCARPS

SPHAGNUM

small, with strongly toothed costa and margins; cells small; on limy rocks with seepage

A small widespread species, locally common in crevices in limy rocks, usually with seepage. *M. marginatum* is similar and grows in the same places. Both are small and somewhat silvery with red stems. *Thomsonii* is separated by its sharper teeth, toothed costa, and smaller cells. A third species, variously called *ambiguum* or *lycopodioides*, is said to have the teeth of *thomsonii* and the cells of *marginatum*. We have not been able to distinguish it clearly from *thomsonii*.

MNIUM THOMSONII

leaves with low rounded teeth and small needle tips; limy rocks with seepage

A small *Mnium* found in the recesses of moist, limy ledges. Widespread in North America, and frequent in the NFR wherever there is suitable habitat. Recognized in the field by the low teeth and the needle tips of the leaves. Confirmed in the lab by the rounded teeth and the rounded corners of the leaf cells. *M. thomsonii*, which often grows with it, has sharper teeth and smaller cells.

MNIUM MARGINATUM

pale silvery unbordered leaves with very small teeth

A small, pale species of moist limy soil; found regularly, usually in small quantities, throughout the NFR. We find it in old quarries, on soil below limy ledges, and on tree bases in fertile woods. The small teeth, silver-green color, and unbordered leaves are good field marks.

QUICK GUIDES TO HABITATS

QUICK GUIDES TO ACROCARPS

QUICK GUIDES TO PLEUROCARPS

QUICK GUIDES TO SPHAGNUM

ACROCARPS

PLEUROCARPS

SPHAGNUM

ONCOPHORUS WAHLENBERGII Slender leaves with clasping bases and long tips, curly when dry; on wood in swamps

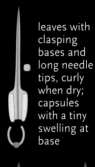

leaves with clasping bases and long needle tips, curly when dry; capsules with a tiny swelling at base

A SLENDER-LEAVED MOSS with curly leaf tips and clasping leaf bases, frequent on well-decayed logs in wet woods or swamps and often mixed with other mosses. Wide ranging in the north and throughout the NFR. The plants are light colored and grow in loose tufts; the leaf tips are needle-like, spread widely when wet, and curl like corkscrews as they dry. *Dicranella heteromalla* and some Dicranums have similar leaves but lack the clasping bases. The tiny bump at the base of the capsule is a good confirmation.

ONCOPHORUS WAHLENBERGII

ORTHOTRICHUM Small mosses in tufts or cushions on bark and rock; leaves lanceolate, their edges mostly recurved; cells papillose; capsules slender and erect; calyptra like a stocking cap*

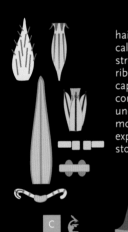

hairy calyptras; strongly ribbed capsules contracted under the mouth; exposed stomates

A LARGE GENUS OF SMALL TUFTED MOSSES, found on trees and rocks, recognized by the erect leaves that are straight when dry and the slender erect capsules that are often surrounded by the leaves. *Ulota* is very similar, and no single character will distinguish all Ulotas from all Orthotrichums. Approximately, anything with smooth or sparsely hairy calyptras and capsules surrounded by leaves or only slightly exserted is an *Orthotrichum*. Anything with strongly curly leaves, long-exserted capsules with long tapering bases, or densely hairy calyptras is a *Ulota*.

We have eight species of *Orthotrichum* in the NFR; most require the microscope. The critical features are the position and ribbing of the capsules and the extent to which guard cells of the stomates are covered by the adjacent cells.

ORTHOTRICHUM SORDIDUM

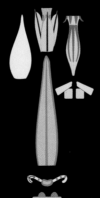

smooth calyptras; capsule ribbed, strongly narrowed below mouth when dry; stomates covered

THE ORTHOTRICHUMS divide by substrate and stomates. *Sordidum*, above, is the common tree-bark species with exposed stomates in the capsule.* It also has hairy calyptras and capsules with eight ribs that are surrounded by the leaves. *Stellatum*, left, is a common tree-bark species with smooth calyptras, covered stomates, and exserted capsules with eight ribs and a strong constriction below the mouth when dry. The peristome teeth in the dry capsules flare widely and look starlike; the ribs of the dry capsules nearly touch.

ORTHOTRICHUM STELLATUM

*The diagrams below the capsules show the guard cells are in green and the adjacent cells in pale blue. *Anomalum* and *sordidum* are the common species with exposed stomates; *stellatum* is the common species with covered stomates. *Speciosum* and *elegans*, not shown, are rare species with exposed stomates and exserted capsules with weak or no ribs. *Ohioense, pumilum*, and *strangulatum* are uncommon species with exposed stomates.

smooth or sparsely hairy calyptras; weakly ribbed capsule with covered stomates and 16 peristome teeth

OHIOENSE is a small tree-bark species that resembles *stellatum.* Its capsules have narrower ribs than *Stellatum,* are less strongly constricted under the mouth, and have 16 peristome teeth. The calyptras may be smooth but are often sparsely hairy. It is widespread on trees in the NFR but not common.*

*ORTHOTRICHUM OHIOENSE**

blunt, flat, straplike leaves, often with reddish-brown brood bodies; tall single papillae

A TINY DISTINCTIVE SPECIES, wide ranging in North America and found throughout the NFR. Fairly common on the trunks of deciduous trees. We look for it on shade trees, particularly maples and apples. Recognized in the field by its small size, olive-green to yellow-green color, and blunt leaf tips. Recognized microscopically by the flat leaf margins and single high papillae, which may be forked at the tip. Almost always sterile.

ORTHOTRICHUM OBTUSIFOLIUM

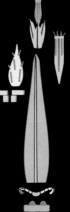

exserted capsules with 8 short ribs alternating with 8 long ones; immersed stomates

THE COMMON *ORTHOTRICHUM* of limy rocks and marble tombstones. Small and dark green, capsules exserted, with eight long ribs, eight partial ribs, and covered stomates. The combination of long and short ribs is unique. *Ulota hutchinsiae,* similar to and sometimes growing with *anomalum,* has hairier calyptras and capsules with reflexed peristome teeth, eight long ribs, and exposed stomates.

ORTHOTRICHUM ANOMALUM

Orthotrichum pumilum is a southern tree-bark species with covered stomates and smooth calyptras, similar to *stellatum* and *ohioense,* and rare in the NFR. It has sharp-pointed leaves with relatively large cells, strongly ribbed capsules, and eight peristome teeth.

ORTHOTRICHUM

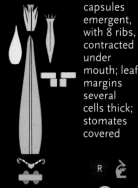

capsules emergent, with 8 ribs, contracted under mouth; leaf margins several cells thick; stomates covered

A RARE OR UNCOMMON *ORTHOTRICHUM* of dry limy rocks, scattered in the southern half of the NFR. Plants often dark green or brown, with sparsely hairy calyptras and emergent capsules that are strongly contracted under the mouth and have 8 ribs. Under the microscope, it has covered stomates, thickened leaf margins, and 16 peristome teeth. *Anomalum*, the common species on limy rocks, has exserted capsules with 8 long ribs and 8 short ones.

ORTHOTRICHUM STRANGULATUM

PALUDELLA SQUARROSA Small moss with hooked leaves in five rows, growing among other mosses in limy fens

elliptical leaves, hooked downward, in five rows

A UNIQUE MOSS, common in fens in the subarctic, much rarer with us. Recognized immediately by the slender stems and five rows of short hooked leaves. May grow in small cushions or scattered among other mosses. Often found as single shoots within clumps of *Sphagnum*.

PALUDELLA SQUARROSA

PARALEUCOBRYUM LONGIFOLIUM Slender silver-green leaves curved to one side, tips filled with costa

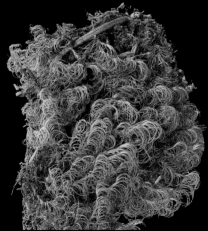

PARALEUCOBRYUM LONGIFOLIUM

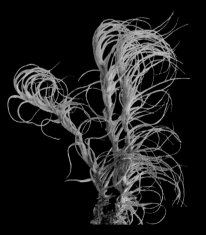

broad flat costa that is mostly empty cells; shaded rocks in cold woods

A LARGE MOSS, common on boulders and ledges in northern woods, either forming large patches or as scattered plants among other species. Light silvery green with slender-tipped leaves curved to one side and a broad flat costa that shows alternating stripes of green and white cells when dry. Closely resembles *Dicranum scoparium*; check for the distinctive costa with a microscope until you know it. *Dicranodontium* also has slender leaves and a broad costa, but its leaves are straighter and more erect, and its costas are not striped.

QUICK GUIDES TO HABITATS

QUICK GUIDES TO ACROCARPS

QUICK GUIDES TO PLEUROCARPS

QUICK GUIDES TO SPHAGNUM

ACROCARPS

PLEUROCARPS

SPHAGNUM

PHILONOTIS Small stringy mosses of wetlands and seeps with light silvery-green sharp-pointed leaves

small sharp-tipped leaves with double teeth; some cells project at lower ends; round capsules, open wet mineral soil

SMALL STRINGY MOSSES of wetlands and seeps, often in large mounds or mats. Light-green or yellow-green; stems red; leaves silvery, with small sharp teeth, the costas often projecting as needle tips; cells rectangular, their ends projecting above and below. *Philonotis fontana* is our common species. *Philonotis marchica* differs, unimpressively, in the amount of toothing and has cells that project at the upper ends only. *Pohlia wahlenbergii*, very similar in the field, has much larger cells whose ends don't project.

PHILONOTIS FONTANA

PHYSCOMITRIUM PYRIFORME Small short-lived moss with translucent spoon-shaped leaves and top-shaped capsules

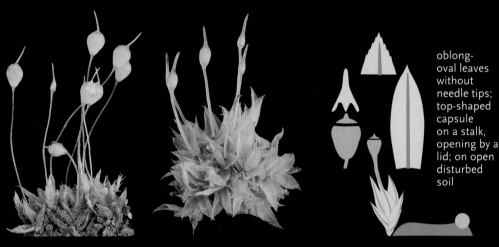

oblong-oval leaves without needle tips; top-shaped capsule on a stalk, opening by a lid; on open disturbed soil

A COMMON MOSS of open disturbed soil, especially common in cultivated fields and on burned soil. Capsules present from late fall through spring. Leaves translucent, pointed, with large thin-walled cells and a weak border. Capsules top shaped, raised above the leaves on a distinct seta, opening by a lid. *Aphanorrhegma*, also common, has almost stalkless capsules with thickened cell corners. *Physcomitrium immersum*, a rare species, is like *Aphanorrhegma* but without the thickened cell corners. The leaves of all three are similar; capsules are required for identification.

PHYSCOMITRIUM PYRIFORME

PLAGIOBRYUM ZIERI Small, very rare moss of limy rocks in the subalpine zone, with silvery leaves and cylindrical shoots

Pale white-green leaves with large, thin-walled cells, in cylindrical shoots; lower stems reddish; long-necked, horizontal capsules

A SMALL RARE MOSS, found in crevices with limy seepage in alpine and subalpine ledges. Looks, except for the elongate capsules, like a larger version of *Bryum argenteum*: pale, white-green, oval leaves, closely overlapped, making cylindrical shoots. Separated from *argenteum* by the larger size, large thin-walled cells, red lower stems, and alpine habitat. *Anomobryum julaceum*, which can grow with *Plagiobryum*, has tapering branches, longer and narrower leaf cells, and no red on the lower stems.

*PLAGIOBRYUM ZIERI**

* The photographs are from specimens collected in Pringle's Ravine, a steep subalpine gully on Mt. Mansfield, by the author in 1998 and Matt Peters in 2017. They come with fine memories.

PLAGIOMNIUM Oval-leaved mosses with bordered leaves, single teeth, and flat arching sterile shoots

single teeth in upper half of the leaf; one seta and capsule per stem

PLAGIOMNIUM CUSPIDATUM

FOREST AND WETLAND MOSSES, with oval leaves, single teeth, and flattened sterile stems, found on soil, rocks, and wood in moist shade. *Cuspidatum,* found throughout much of North America, is the common species of deciduous woods. It is recognized easily by the relatively small leaves that are only toothed above the middle. It tends to be replaced by the toothier *ciliare* or *ellipticum* in swampy woods.*

leaves with sharp tips, toothed to base; long teeth with 2-4 cells, blunt at tips

PLAGIOMNIUM CILIARE

THE SECOND-COMMONEST FOREST *PLAGIOMNIUM,* after *cuspidatum,* found throughout the NFR on wet soil, logs, rocks, and tree bases. The long multi-celled teeth, extending to the base of the leaf, leaves with small needle tips, and decurrent bases are usually sufficient to identify it. *P. ellipticum,* a swamp species that was formerly combined with *ciliare,* has shorter teeth, non-decurrent bases, and multiple setae.

non-decurrent leaves, often rounded at the tip, toothed to base with 1-2 celled, blunt teeth; plants mostly sterile; in swampy woods

PLAGIOMNIUM ELLIPTICUM

A THIRD COMMON, WIDE-RANGING SPECIES, related to *ciliare* but with shorter teeth and non-decurrent leaves that tend to be rounded at the tips. Typically found in hummocks in swamps, on stream banks, and in seeps. It is quite variable and so hard to confirm. A useful microscopic feature is the large leaf cells near the costa that contrast with smaller marginal cells.

* *Plagiomnium drummondii* is a rare boreal species that is like *cuspidatum* but with longer teeth, bigger cells and multiple setae. We have never seen it.

QUICK GUIDES TO HABITATS

QUICK GUIDES TO ACROCARPS

QUICK GUIDES TO PLEUROCARPS

QUICK GUIDES TO SPHAGNUM

ACROCARPS

PLEUROCARPS

SPHAGNUM

very large; leaves with decurrent bases and sharp teeth; multiple setae

OUR LARGEST PLAGIOMNIUM and one of our most wide ranging, found from the arctic to the warm temperate zone. Uncommon but locally abundant in the NFR in woods, on limy ledges, and old quarries. Seems to like moist sheltered limy sites. Recognized by the large size, multiple setae, decurrent base, and sharp 1-celled teeth.

PLAGIOMNIUM MEDIUM

creeping shoots; rounded nondecurrent leaves with blunt teeth and cells thickened at the corners; multiple setae

A WIDE-RANGING SPECIES, reported from the arctic to the tropics. Uncommon in the NFR where it is found on soil and ledges where there is limy seepage. It is a medium-sized species that grows flat and roots along the stems. It resembles *ellipticum* in its non-decurrent leaves with rounded or squared-off tips and short teeth. The best characters are the small, nonpitted cells that are strongly thickened at the corners and, when fertile, the multiple setae and long beaks on the capsules.

PLAGIOMNIUM ROSTRATUM

PLAGIOPUS OEDERIANA C M Slender arching double-toothed leaves in three rows; round capsules; always in limy seepage

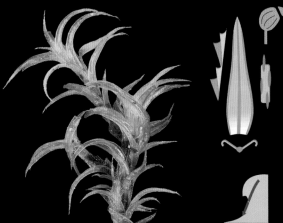

lanceolate double-toothed leaves with recurved edges and elongate papillae; round capsules

A MEDIUM-SIZED MOSS that makes loose tufts or cushions on rocks with limy seepage. Related to *Bartramia*, and, like it, with slender, doubly toothed leaves and round capsules. Resembles *Amphidium*, which grows in similar places and has similar elongate papillae that cross the cell walls. *Amphidium* has no teeth, and neither *Amphidium* nor *Bartramia* has three-rowed leaves.

PLAGIOPUS OEDERIANA

PLEURIDIUM SUBULATUM Spring ephemeral of open soil; round capsules surrounded by slender leaves with needle tips

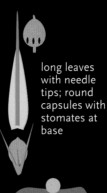

long leaves with needle tips; round capsules with stomates at base

SMALL SHORT-LIVED MOSS of cultivated ground and moist open soil, recognized by the needle-tipped leaves, hooded calyptra, and round capsules with stomates at base. Seems uncommon but can be locally abundant. *Bruchia flexuosa*, a southern species, uncommon with us, has similar leaves but an elongate capsule with a neck.

PLEURIDIUM SUBULATUM

POGONATUM Mosses with stiff opaque leaves with photosynthetic lamellae (ridges); leaves mostly under 1 cm long; capsules cylindrical, without angles; terminal cells of the lamellae papillose

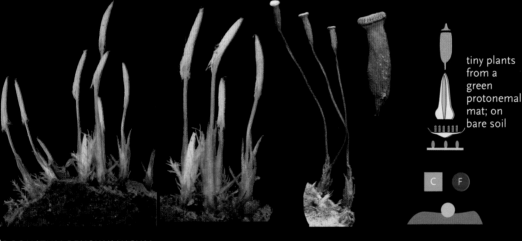

tiny plants from a green protonemal mat; on bare soil

MOSSES WITH PHOTOSYNTHETIC LAMELLAE with papillose terminal cells and cylindrical capsules. We have 0 species that grows from a persistent protonema and two leafy species resembling Polytrichums. Compare *Polytrichastrum alpinum*, which also has papillose terminal cells but is larger and has narrower leaves.

Pogonatum pensilvanicum, which grows from a mat of felty protonema, is common on bare soil, especially moist banks. The mat of protonema may be extensive; the gametophytes are tiny and capsules large.

POGONATUM PENSILVANICUM

broad blue-green leaves, on short stems; cylindrical capsules; terminal cells of lamellae flat topped and papillose; leaf sheaths without a border

A SMALL BLUE-GREEN SPECIES with short, relatively broad leaves, looking like small palms. The best diagnostic features are the flat tops of the terminal cells of the lamellae and the absence of a border on the sheaths. Common on open soil in the alpine zone, occasional at lower elevations, especially on the northern coast. Very similar to *Pogonatum urnigerum* and often grows with it. Sometimes *urnigerum* is taller or toothier. Sometimes it is not, and only the microscope knows for sure.

POGONATUM DENTATUM

QUICK GUIDES TO HABITATS

QUICK GUIDES TO ACROCARPS

QUICK GUIDES TO PLEUROCARPS

QUICK GUIDES TO SPHAGNUM

ACROCARPS

PLEUROCARPS

SPHAGNUM

broad blue-green leaves, on short stems; cylindrical capsules; terminal cells of lamellae with round tops; leaf sheaths with a border of long cells

SIMILAR TO *DENTATUM* and, like it, an arctic-alpine species common at high elevations and occasional at low ones. Often taller and more branched than *dentatum*, with narrower leaves and stronger teeth; sometimes low and indistinguishable from it. The most reliable diagnostic features are microscopic: the sheaths of *urnigerum* have a border of clear thin-walled cells, and the terminal cells of the lamellae have rounded tops.

POGONATUM URNIGERUM

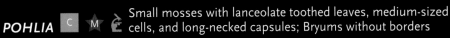

POHLIA Small mosses with lanceolate toothed leaves, medium-sized cells, and long-necked capsules; Bryums without borders

thick-walled cells; brood bodies absent; capsules with well developed inner peristome segments and cilia

A COMMON GENUS of small weedy mosses that grow in tufts and have lanceolate leaves with low teeth and medium-sized oblong-rhombic cells. The capsules resemble those of *Bryum* but the leaves are unbordered. Red stems, various sorts of brood bodies, and leaves with a metallic sheen are common in the genus. Accurate identification of the species usually requires a microscope.* *Pohlia nutans*, which makes dense mats on banks and in rock crevices, is the commonest species. Other than being a *Pohlia*, it is not distinctive. We assume that all ordinary-looking *Pohlias* are *nutans* and then check the cells and peristome teeth if we need to be sure. Since it is so common, often we don't.

POHLIA NUTANS

pale-green with a strong metallic iridescence; long, thin-walled cells

A SMALL SPECIES, found in crevices in rocks and moist shaded soil, especially along streams or where there is limy seepage. Identifiable in the field from the pale color and metallic iridescence. *Pohlia nutans* grows in similar places and has some iridescence as well. In the field it is yellower green and grows in denser clumps than *cruda*. Under the microscope its cells are shorter and thicker walled.

POHLIA CRUDA

* Besides the species shown here, *Pohlia andalusica, elongata, lescuriana,* and *proligera* occur rarely in the NFR. *Andalusica* and *proligera* are bulblet-bearing species, differing from *annotina* in the shape of the bulblets. *Elongata* and *lescuriana* resemble *nutans* but differ in the details of the capsules and peristome. We have looked for them for many years but have never seen them.

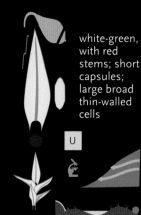

POHLIA WAHLENBERGII

Pohlia   *Philonotis*

white-green, with red stems; short capsules; large broad thin-walled cells

WHITE-GREEN PLANTS with red stems in clumps in wetlands and on wet banks; capsules short, cells thin walled and wide. Very similar in the field to *Philonotis fontana,* which grows in similar places. Both have red stems, lanceolate leaves, a white-green color in the spring, and a bit of metallic sheen. *Philonotis* has longer, more needlelike leaf tips and narrower cells that protrude at their ends.

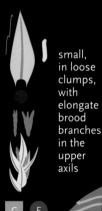

POHLIA ANNOTINA

small, in loose clumps, with elongate brood branches in the upper axils

A SMALL, PALE-GREEN MOSS found on road banks, disturbed soil, and occasionally in seepage in cracks of rocks. Slender and pale; needs to be looked for. Distinguished from other *Pohlia*s by the tiny brood branches in the upper axils. The branches have three or more leaf primordia at their tips. The rare *Pohlia proligera* is similar but has fewer primordia. Often grows with *Pohlia bulbifera,* which has ball-shaped gemmae instead of brood branches.

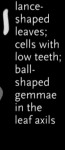

POHLIA BULBIFERA

lance-shaped leaves; cells with low teeth; ball-shaped gemmae in the leaf axils

SMALL SLENDER GLOSSY MOSSES with round gemmae (consisting of a stem with a couple of leaves rolled in over its tip) in the leaf axils. Found mostly in moist, open disturbed soil: road edges, gravel pits, pond shores, etc. The ball-shaped gemmae tucked in close to the stem are the diagnostic character. The rare *Pohlia andalusica* has top-shaped gemmae; *annotina* and the rare *proligera* have elongate brood branches.

QUICK GUIDES TO HABITATS

QUICK GUIDES TO ACROCARPS

QUICK GUIDES TO PLEUROCARPS

QUICK GUIDES TO SPHAGNUM

ACROCARPS

PLEUROCARPS

SPHAGNUM

POLYTRICHASTRUM ALPINUM C M Ê

Opaque slender leaves with photosynthetic lamellae, resembling spruce needles; leaf blades slender; capsules cylindrical; top cells of lamellae papillose

Photo-synthetic lamellae on slender leaves; capsules cylindrical; top cells of lamellae papillose

A COMMON TALL MOSS of boulders and ledges in cool moist woods. Vegetatively similar to *Polytrichum commune* and *pallidisetum* but having cylindrical capsules (no ridges or angles) and papillose cells on the edge of the lamellae. *Pogonatum dentatum* and *urnigerum* have similar papillose cells but are shorter, with shorter leaves with broader blades, and usually more blue green.

POLYTRICHASTRUM ALPINUM

POLYTRICHUM C M Ê Opaque needlelike leaves with lamellae; capsules angled; terminal cells of lamellae not papillose

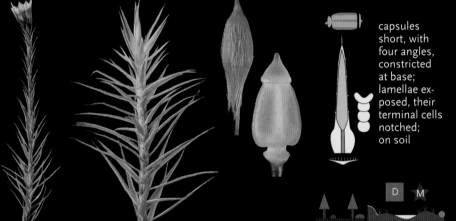

capsules short, with four angles, constricted at base; lamellae exposed, their terminal cells notched; on soil

D M

LARGE OR MEDIUM MOSSES with opaque needle-like leaves covered by vertical photosynthetic lamellae (ridges). Our species have angled capsules and smooth rather than papillose cells on the upper edge of the lamellae, distinguishing them from the related, and often similar-appearing, genera *Pogonatum* and *Polytrichastrum*. They grow on soil and rock in many different habitats: forests, swamps, bogs, meadows, and barrens.

Commune is large, common and widespread; its best field mark is the short capsule with a constricted base. Sterile plants resemble *P. pallidisetum*, *P. ohioense*, and *Polytrichastrum alpinum*. They are identified microscopically by the smooth, consistently notched cells at the edge of the lamellae.

POLYTRICHUM COMMUNE

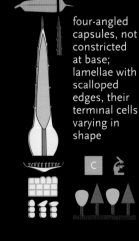

four-angled capsules, not constricted at base; lamellae with scalloped edges, their terminal cells varying in shape

C Ê

THE COMMON *POLYTRICHUM* of moist northern forests and swamps. Typically tall and graceful, with four-angled capsules that are not constricted at base and lamellae with intermittently scalloped edges. *Ohioense,* the common forest-floor species in the oak zone, is shorter and has thick-walled, flattened cells on the lamellae edge. *Formosum* and *longisetum,* rare species of northern conifer forests, have narrow (taller than wide) cells on the edges of the lamellae.

*POLYTRICHUM PALLIDISETUM**

* *Polytrichum longisetum, pallidisetum,* and *ohioense* were moved to *Polytrichastrum* in the *Flora of North America.* This was a mistake

POLYTRICHUM LONGISETUM

four-angled capsules, not constricted at base; lamellae with flat edges; terminal cells that are taller than wide. Leaf margins often wide, inrolled on dry plants.

R

A NORTHERN *POLYTRICHUM* of cool forests, wetlands, and tundra with a scattered distribution in the NFR. Very similar in the field to *pallidisetum*, from which it is separated microscopically by the lamellae, which have straight edges without scalloping and terminal cells that average a bit higher than wide. *Polytrichum formosum* (not shown) is a similar species, also rare, separated by its larger capsules, somewhat narrower leaf margins and somewhat longer cells.

POLYTRICHUM OHIOENSE

four-angled capsules, not constricted at base; lamellae with flat edges and thick-walled terminal cells with flat tops

C

THE COMMON *POLYTRICHUM* of forest floors in the oak zone: plants relatively short, lamellae exposed, capsules angled but not constricted at base, edge cells of the lamellae thick walled and flattened. *Commune,* which can grow with it, is separated by the short capsules with constricted bases and notched cells on the edges of the lamellae. *Pallidisetum,* though usually taller and more northern, can only be separated by the thin-walled cells on the lamellae edges.

POLYTRICHUM JUNIPERINUM   POLYTRICHUM STRICTUM

edges of leaves inrolled, covering the lamellae; leaves with and dark needle tips

D   M

TWO COMMON SPECIES of open habitats: dry forests, barrens, bogs and heaths, sand plains, and alpine tundra. Recognized by the combination of inrolled leaf edges that cover the lamellae, and the dark needle tips. *Piliferum,* which also has covered lamellae, has soft white hair-points. *Commune, pallidisetum,* and *ohioense* have exposed lamellae and teeth along the edges. *Strictum* is a long narrow relative of *juniperinum,* common on bog hummocks. It intergrades with *juniperinum* and is probably best regarded as an ecological form of it.

QUICK GUIDES TO HABITATS

QUICK GUIDES TO ACROCARPS

QUICK GUIDES TO PLEUROCARPS

QUICK GUIDES TO SPHAGNUM

ACROCARPS

PLEUROCARPS

SPHAGNUM

leaf edges inrolled, covering lamellae; leaves ending in a white hair point

D   M

OUR SMALLEST *POLYTRICHUM*, immediately recognizable by the treelike growth form, thick leaves with inrolled edges and white hair-tips. Forms low groves in dry, rocky or sandy places: barrens, ledge crests, sandy roadsides, usually in full sun. Hairlike white tips are found in other genera, particularly *Grimmia* and *Racomitrium*. They have thinner leaves and form clumps rather than groves.

POLYTRICHUM PILIFERUM

PSEUDOBRYUM CINCLIDIOIDES   C   M   Large moss resembling *Plagiomnium* or *Rhizomnium* but with only a weak border and a few tiny teeth

large dark-green plants with oblong leaves with weak borders and a few tiny teeth; cells large, diamond shaped, with pointed ends

A DISTINCTIVE MOSS of wet mucky or organic soils: swamps, fens, woodland pools that dry in the summer. Distinguished by the large broad leaves with a weak border of elongate cells and a few small teeth. The cells are large and diamond shaped, in oblique rows, and the border is not thickened. *Rhizomnium appalachianum* and *Plagiomnium medium* look generally similar but have much stronger borders and, in *Plagiomnium*, more distinct teeth.

PSEUDOBRYUM CINCLIDIOIDES, with *Sphagnum squarrosum*

RACOMITRIUM✳   C   M   Dark shaggy mosses, with many short side branches, on rocks and soil; leaves often with white needle tips; cells with wavy sides; capsules with a beak.

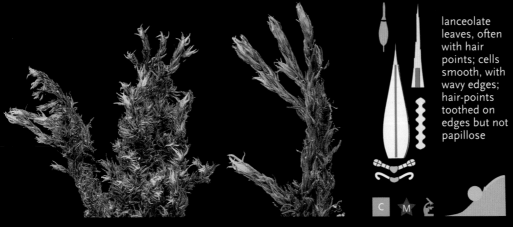

lanceolate leaves, often with hair points; cells smooth, with wavy edges; hair-points toothed on edges but not papillose

C   M

DARK MOSSES WITH LANCEOLATE LEAVES, commonly making cushions or patches on rocks, often much branched, often with white needle tips. Separated from all our other genera by the extremely wavy walls to the cells. Related to *Grimmia* and *Schistidium* and often growing with them; usually separable in the field by the abundant side branches of *Racomitrium*. The *venustum* group *(venustum, affine, sudeticum,* and *microcarpon)* is common on rocks that are intermittently wet, especially on exposed cliffs and summits. They have hair-points with teeth but no papillae.

RACOMITRIUM VENUSTUM GROUP

✳ European botanists separate three genera, *Bucklandiella, Codriophorus,* and *Niphotrichum,* from *Racomitrium.* Our friend Nomenclator says "Names are cheap over there." The *venustum* group is called *Racomitrium heterostichum* in older books; the species are distinguished by small differences in the structure of the leaves.

RACOMITRIUM

on wet
rocks;
tips
broadly
rounded;
costa
ends
near tip

A COMMON SPECIES of shaded rocks in the splash-zone of streams, with oblong leaves with rounded tips and recurved edges. Under the microscope the cells are typical wavy-edged *Racomitrium* cells and the costa nearly reaches the tip. *Schistidium rivulare* and *agassizii* look similar and grow in similar habitats. In the field, *rivulare* has sharp-pointed tips and *agassizii* has flat rather than keeled tips. The three are close and checking with the microscope is necessary.

RACOMITRIUM ACICULARE

on wet
rocks; tips
inrolled
and
pointed;
costa
ends well
below tip

A NEW SPECIES, created from *aciculare* in 1999. It grows in the same habitats as *aciculare* and looks identical superficially. The only differences are that the leaf tips are narrower and the costa shorter. We include it here because it seems common where we work; we are not really impressed by the differences and can't vouch for their consistency.

RACOMITRIUM ADUNCOIDES

leaves grey
green,*
densely
papillose,
with papillose
hair-points;
on exposed
soil, boul-
ders, and
ledges

AN UNCOMMON NORTHERN SPECIES, found on bare sand, gravel, and exposed rocks. Widely distributed but not common anywhere in the NFR. Medium sized, olive or yellow green, whitened when dry, with white papillose leaf tips that are finely toothed. Under the microscope the long papillose needle tips and the tall papillae over the leaf cells are unique. *Lanuginosum* is similar but bigger and woollier, with longer, more sharply toothed tips.

RACOMITRIUM CANESCENS

* Our pictures are from a specimen that had yellowed.

QUICK GUIDES TO HABITATS

QUICK GUIDES TO ACROCARPS

QUICK GUIDES TO PLEUROCARPS

QUICK GUIDES TO SPHAGNUM

ACROCARPS

PLEUROCARPS

SPHAGNUM

leaves tapering into a long tip, rounded at the extreme tip, without hair-points; costa not reaching tip; cells papillose, in one layer

A NORTHERN SPECIES, uncommon in the NFR. Found on wet rocks by lakes and streams at low elevations and on exposed rocks in the alpine zone. Locally common in the Adirondack High Peaks and likely elsewhere at high elevations. Generally similar to the *venustum* group in the field but with long narrow leaf tips without hair-points. Confirmed microscopically by the papillose leaf cells, relatively short costa, and one-layered margins.

RACOMITRIUM FASCICULARE

large gray-green moss of barrens and summits; leaves with long, strongly toothed, decurrent hair-points

A BIG SHOWY MOSS that covers large areas in the far north. It is found on rocky barrens in the arctic and sub-arctic, and on limestone and serpentine southwards. It is rare in the mountains and along the northern coast with us; its most spectacular NFR occurrence may be on the serpentine tablelands of Mont Albert. The plants are grey-green, with long white hair-points that make them look woolly. The best diagnostic characters are the long hair-points, which are papillose and strongly toothed. *Canescens* has a similar color but weakly toothed hair-points and higher papillae on the cells.

RACOMITRIUM LANUGINOSUM

RHABDOWEISIA CRISPATA Small moss with slender leaves and erect furrowed capsules, making mats on rock

linear leaves with short square cells; erect furrowed capsules

A SMALL SPECIES, reasonably common on shaded moist ledges, often in crevices or recesses, usually on acid rock. Found throughout the NFR and down the Appalachians. Leaves linear, fairly flat, with smooth cells and a few separated teeth near the tips. Capsules cup shaped, furrowed when dry, with a peristome of undivided teeth. *Gymnostomum* and *Hymenostylium* look similar in the field; their leaves are papillose and their capsules lack peristomes and furrows.

RHABDOWEISIA CRISPATA

RHIZOMNIUM  D M Erect mosses of wet places; broadly oval untoothed leaves with strongly thickened borders

large oval strongly bordered leaves without teeth; abundant rhizoids on stem and leaf bases

COMMON ROUND-LEAVED MOSSES with untoothed, strongly bordered leaves, always in wet shady places. *Plagiomnium* has similar leaves but has teeth and flattened sterile shoots.

Appalachianum is a large common species of pools in swamps, making big colonies. It has leaves over 6 mm long and rhizoids on both the stem and bases of the leaves. *Pseudobryum cinclidioides* looks similar and grows in the same sort of places but has unbordered leaves.*

RHIZOMNIUM APPALACHIANUM

small oval strongly bordered leaves without teeth; plants from a mat of brown rhizoids

A SMALL COMMON SPECIES found in wet shaded places: ledges, rocks in streams, banks with seepage, wet shaded soil. Oval bordered leaves without teeth, silvery when young, dark green or reddish when old, arising from a mat of brown rhizoids; stems furry only at base. *Appalachianum*, our other common species, is much larger and has rhizoids on the stem but no rhizoidal mat.

RHIZOMNIUM PUNCTATUM   male plants

RHODOBRYUM ONTARIENSE C M Treelike moss with a rosette of oboval leaves at the top of the stem

large oboval leaves in a rosette at the top of the stem; in colonies in fertile woods

A LARGE MOSS of fertile soils, forming groves on tree bases, logs, and limy rocks. Immediately recognizable from the size, obovate leaves, and growth form. Some *Bryums* are similar but much smaller.

RHODOBRYUM ONTARIENSE (with *Thuidium delicatulum*)

* *Rhizomnium magnifolium* and *pseudopunctatum* are northern and western species resembling *appalachianum* that must be distinguished with the microscope. Both are local near the western Great Lakes and very rare in the eastern NFR.

QUICK GUIDES TO HABITATS

QUICK GUIDES TO ACROCARPS

QUICK GUIDES TO PLEUROCARPS

QUICK GUIDES TO SPHAGNUM

ACROCARPS

PLEUROCARPS

SPHAGNUM

SAELANIA GLAUCESCENS ^F ^U Small moss with slender pale-green leaves; leaves and stems with a blue-white, spider-webby coating

slender pale-green leaves; square cells; blue-white filaments, like cobwebs, on stem and leaves

SAELANIA GLAUCESCENS

A SMALL DISTINCTIVE MOSS found in crevices in rocks or in clumps in rocky barrens, often with calcareous seepage. Leaves pale green, linear, with small square cells and some teeth near the tip; stem and lower leaves covered with blue-white, spider-webby filaments. The filaments, which are the best field character, are said to be made by the plant and not fungi or spiders. How the plant makes them, no one knows.

SCHISTIDIUM ^C Dark narrow-leaved mosses, resembling *Grimmia* but in looser clumps; leaves with at most short needle tips; capsules surrounded by enlarged perichaetial leaves; central column of capsule remains attached to lid.

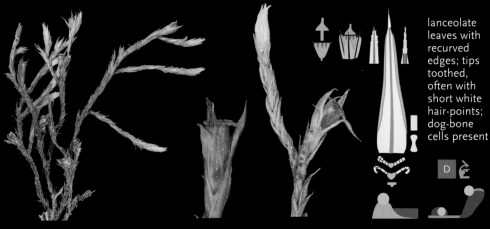

lanceolate leaves with recurved edges; tips toothed, often with short white hair-points; dog-bone cells present

*SCHISTIDIUM APOCARPUM GROUP**

A COMPLEX GENUS, treated minimally here. Dark mosses, growing in loose patches on wet or dry rocks; leaves lanceolate, often with recurved edges and wavy-edged cells; capsules short and cup shaped, the central column falling with the lid. Close to and formerly a part of *Grimmia*. The looser clumps, immersed capsules, and missing central columns are the best ways of separating them.

The *apocarpum* group is common on both acid and limy rocks, often where there is at least intermittent seepage. They have short hair-points, usually a few teeth, and often have papillae on the lower side of the costa near the tip; fertile plants have oblong capsules, narrower than those of *rivulare*.

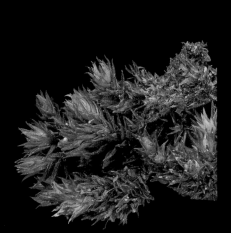

Pointed leaves without hair-points; capsules short; costa not papilllose; rocks by streams

A DARK SHAGGY MOSS found on rocks in and near streams. Generally like the *apocarpum* group, but the leaves without hair-points, the capsules shorter, the costa not papillose, and the cells not particularly wavy-edged. Plants intermediate between the two are not hard to find; *Schistidium* is like that.

SCHISTIDIUM RIVULARE

* Including *Schistidium apocarpum* and a group of close relatives, many of them recently separated: *crassithecium, lancifolium, lilliputianum, papillosum, pulchrum, viride* ... The distinctions are microscopic. Their characters, ecology, and abundance are not well known.

SCHISTIDIUM

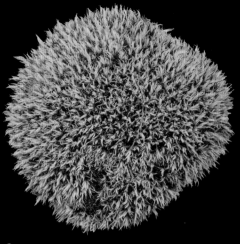

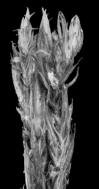

blunt concave linear leaves, two-layers thick near the tip; on rocks by salt water

AN UNCOMMON SPECIES making small, dark cushions on seacoast rocks. Usually above the range of storm tides but still in the spray zone. Leaves linear, concave, two-layered above, with blunt tips without needle points; costas in cross-section with a layer of enlarged cells ("guide cells") across the center.* The rare *Ulota phyllantha* grows in similar places.

SCHISTIDIUM MARITIMUM

SCHISTOSTEGA PENNATA [U] [F] Tiny pale-green flattened fronds growing from a reflective protonema in recesses under ledges and boulders

reflective cells of protonema

tiny moss of dark recesses; reflective protonema with bubble cells; tiny fernlike fronds; short capsules

crevices in ledges, under boulders

A SMALL, UNCOMMON MOSS that lives in dark recesses in and under rocks, also sometimes in old barn foundations and under stumps. The protonema has reflective chloroplasts that focus light back at you. The flattened fronds resemble those of *Fissidens*, but the leaves do not have pockets.

SCHISTOSTEGA PENNATA

SELIGERIA CAMPYLOPODA [R] 🔬 Tiny mosses forming dark mats on limy rock; capsules nodding, costa doesn't fill leaf tip

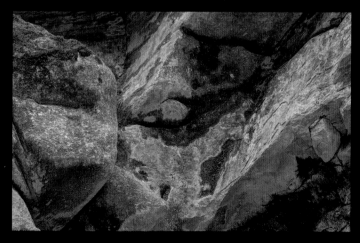

tiny plants with linear leaves and short capsules, in thin, algae-like patches on limy rocks

A GENUS OF TINY MOSSES restricted to limy rocks. Five species occur in our area. All are rare and hard to find. The plants are only a millimeter or two high and barely noticeable until they fruit. The leaves are slender, the cells are smooth and rectangular, the costas are strong and sometimes excurrent, the alar cells are undifferentiated, and the capsules short and sometimes nodding. All seem to be rare, even in the right habitats.

Camplyopoda, shown here, has a slender costa that doesn't fill the tip and setae that are curved when wet.

SELIGERIA CAMPLYOPODA, in thin mats, mixed with *Fissidens subbasilaris*

*A blunt-leaved *Schistidium* with flat leaves from rocks in fresh-water streams may be the rare *Schistidium agassizii*.

QUICK GUIDES TO HABITATS

QUICK GUIDES TO ACROCARPS

QUICK GUIDES TO PLEUROCARPS

QUICK GUIDES TO SPHAGNUM

ACROCARPS

PLEUROCARPS

SPHAGNUM

SPLACHNUM F R Small mosses on dung; leaves sharply toothed; capsules with long swollen necks

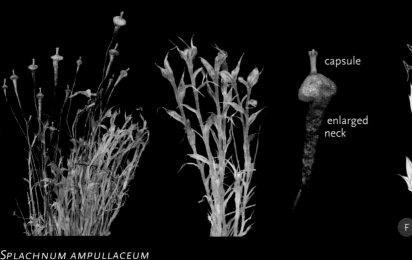

capsule

enlarged neck

pale, sharply toothed leaves with large thin-walled cells; neck of capsule fat and round; on dung of herbivores

F R

THE DUNG MOSSES are small, northern, and rare. Most have small basal leaves with thin-walled cells and large capsules on long setae. The spores are insect-dispersed and the capsules are often colored and scented. *Splachnum*, our commonest genus, has sharply toothed leaves and capsules with greatly enlarged necks. We have three northern-forest species. *Ampullaceum*, the only one that is widespread in the NFR, grows on moose dung in swamps and beaver latrines on shores. The capsules have top-shaped necks and are pink or red when mature.*

SPLACHNUM AMPULLACEUM

SYNTRICHIA RURALIS F R Papillose mosses with oblong light-green leaves with long hair-points; often on limy rocks

oblong papillose leaves with long, toothed points and recurved edges; peristome twisted

SYNTRICHIA is a large genus, common in the west but rare in the east. The distinguishing features are the oblong-oval leaves with long hair-points and the capsules with a peristomes of twisted threads. *Ruralis* is the only widespread species in the NFR. It is medium sized, grows on limy rocks in the open, and has recurved edges and a toothed hair-point. *Encalypta rhaptocarpa* looks similar but grows on wetter ledges and has flat edges and an untoothed point.

SYNTRICHIA RURALIS

TETRAPHIS D M Small mosses with oval leaves and gemmae cups; slender capsules; peristomes of four teeth; make big shaggy mats on logs and stumps

gemmae cup

oval leaves without teeth or borders; gemmae cups common; capsules with four teeth

A SMALL COMMON MOSS, growing on well decayed wood in forests. A unique genus: oval leaves, somewhat like a *Plagiomnium* without a border or teeth; gemmae cups at the tips of sterile stems; long capsules, peristomes of four teeth. *Pellucida* is our common species. *Geniculata*, with narrower leaves and a distinctive bent seta, is a rare species found in conifer forests near the coast in eastern Maine and the Maritimes.

TETRAPHIS PELLUCIDA

* Two other species of *Splachnum, rubrum* and *luteum*, occur rarely in the NFR. Both have the necks of their capsules expanded into parasols. In *rubrum* the parasol is dark red; in *luteum* it is yellow.

TETRAPLODON Small mosses with erect capsules with long necks, found on bones, pellets, and carnivore dung

long neck

oval or oboval leaves with long tips and large thin-walled cells; capsules with long necks, red at maturity

A RARE NORTHERN GENUS of dung mosses: oval leaves with long needle tips and capsules with long necks that are slightly enlarged. The capsules are green when young and dark red at maturity. Both of our species are found on carnivore dung and weathered bones and antlers, most often in alpine zones and tundra. *Angustatus* is tall, with toothed leaves and capsules on short stalks; *mnioides* is budlike, with untoothed leaves and capsules on long stalks.*

TETRAPLODON ANGUSTATUS   *TETRAPLODON MNIOIDES*

TIMMIA MEGAPOLITANA 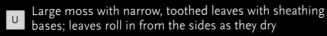 Large moss with narrow, toothed leaves with sheathing bases; leaves roll in from the sides as they dry

long leaves, clasping at their bases, toothed above; cells bulging; calyptra points up

A LARGE, DISTINCTIVE SPECIES associated with limy seepage. Leaves lanceolate, sharply toothed from a sheathing base, rolling into tubes as they dry. Cells round and bulging. Capsules cylindrical, calyptras standing up straight behind the capsules. Locally common on shaded, moist limy soil: stream banks, seepage wetlands below banks, wet ledges with limy seepage.

TIMMIA MEGAPOLITANA

TORTELLA 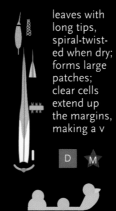 Narrow-leaved papillose mosses of limy places; clear cells at leaf base extend up leaf margins, making a v

leaves with long tips, spiral-twisted when dry; forms large patches; clear cells extend up the margins, making a v

LIMESTONE MOSSES, usually yellow-green, the leaves multipapillose and opaque above, with clear cells extending up the margins and forming a v. *Tortuosa*, which forms large patches and mounds on open ledges and limy soil, is our largest and commonest species. It has leaves with needle tips that are s-curved when wet and twisted like corkscrews when dry.

TORTELLA TORTUOSA

* *Tayloria serrata*, a third species of dung moss, resembles *Tetraplodon* but has leaves with shorter tips and capsules with narrower necks. It is a northern and western species, very rare in the NFR.

QUICK GUIDES TO HABITATS

QUICK GUIDES TO ACROCARPS

QUICK GUIDES TO PLEUROCARPS

QUICK GUIDES TO SPHAGNUM

ACROCARPS

PLEUROCARPS

SPHAGNUM

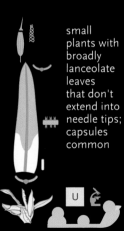

small plants with broadly lanceolate leaves that don't extend into needle tips; capsules common

A SMALL *TORTELLA* without the long corkscrew tips of *tortuosa*. Small plants, often in dense cushions, found on limy rocks and soil, either in the open, or in crevices or under shrubs. Almost always fertile. *Inclinata* is about the same size and also has short-tipped leaves; it has hooded tips to the leaves, makes very dense cushions, and almost never fruits.

TORTELLA HUMILIS

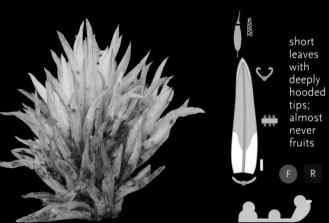

short leaves with deeply hooded tips; almost never fruits

A SMALL, RARE SPECIES of open limy cliffs, flat-rock barrens, and soil, recognized by the relatively broad short leaves that are inrolled and hooded at their tips. It is a wide-ranging northern and western species, rare in the east. It grows in dense bright-colored cushions, always sterile, on open, intermittently wet, limestone ledges and alvars.

TORTELLA INCLINATA

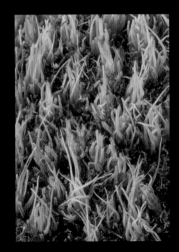

Leaves slender, two-layered above, erect and not much curved when dry, ending in thickened needle tips which break off

A WIDE-RANGING SPECIES, found from Greenland and the North American arctic islands to New Zealand and Antarctica. Scattered and uncommon in the NFR, growing in a variety of intermittently wet, limy places. We find it in small quantities in the crevices of limy cliffs, often in seepage. Farther north it makes turfs in limy barrens. Recognized by the straight erect leaves with thickened tips that break off. The leaf blade is two-cells thick just below the tip.

TORTELLA FRAGILIS

TREMATODON [U] (F) Small mosses of disturbed soil; leaves slender, long tipped and arching; capsules long necked

leaves untoothed, abruptly contracted to the long tip; neck of capsule about as long as the urn

AN INCONSPICUOUS MOSS of northeast North America, found throughout the NFR on open, disturbed mineral soil. Probably common but almost impossible to recognize without capsules. Roadsides, gravel pits, and ditches are common habitats. Leaves with an oval base, abruptly contracted to a narrow tip; costa expanding to fill the tip. Capsules long necked, with a small bump at the base of the neck. The leaves are much like *Ditrichum* or *Dicranella*, through more abruptly contracted to the tip; the long-necked capsules are distinctive.

TREMATODON AMBIGUUS

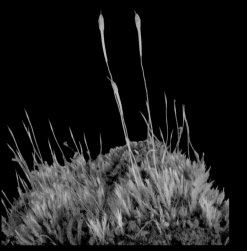

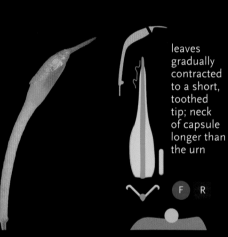

leaves gradually contracted to a short, toothed tip; neck of capsule longer than the urn

A SOUTHERN SPECIES, rare in the NFR, that grows the same habitats as *ambiguus* but is much rarer. Its marks are the relatively short tips of the leaves and the extremely long necks of the capsules. We sometimes find the two species growing together.

TREMATODON LONGICOLLIS

TRICHOSTOMUM TENUIROSTRE [C] (F) 🔬 Small papillose moss with flat leaf edges; tips of the leaves often broken

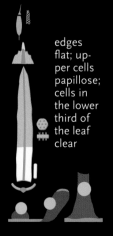

edges flat; upper cells papillose; cells in the lower third of the leaf clear

A SLENDER-LEAVED PAPILLOSE MOSS of tree bases and rocks, similar to *Gymnostomum, Hymenostylium, Tortella,* and others, often recognizable in the field by the broken leaf tips. Wide ranging in temperate North America and common in the southern half of the NFR. Confirmation requires the microscope: the multipapillose upper cells, large clear area at the base, and plane edges are good characters.

TRICHOSTOMUM TENUIROSTRE

QUICK GUIDES TO HABITATS

QUICK GUIDES TO ACROCARPS

QUICK GUIDES TO PLEUROCARPS

QUICK GUIDES TO SPHAGNUM

ACROCARPS

PLEUROCARPS

SPHAGNUM

Narrow-leaved dark mosses making cushions on tree bark or rocks, close to *Orthotrichum* but differing, variously, in the more exserted capsules, hairier calyptras, or more crisped leaves. Sometimes not differing much at all.

hairy
calyptras;
exserted
capsules;
leaves
curly
when dry

C

M

A **common genus** of tufted mosses of bark and rocks, close to *Orthotrichum* and often growing with it. Both have dark lanceolate leaves, mostly with recurved edges; elongate, ridged capsules; and caplike calyptras. The capsules of *Ulota* are always exserted and taper gradually into a long neck; the lower leaf cells often angle outwards from the costa. The capsules of *Orthotrichum* never have long-tapering bases and are sometimes surrounded by the leaves; its lower leaf cells don't flare. *Ulota crispa* is our commonest species, and in fact the commonest tufted moss on tree bark. It is recognized by the leaves that are strongly curly when dry and exserted capsules whose mouths stay open when they dry.

JLOTA CRISPA

leaves slightly
curly when
dry; capsules
with puckered
mouths, wet
or dry

C

M

A **common northern species**: larger, more montane and never as common as *Ulota crispa*, though often found with it. Often taller, less curly, and in loose clumps than *crispa*; if fertile with distinctive club-shaped capsules with puckered mouths. Care is needed: *coarctata* capsules pucker dry and wet; *crispa* capsules flare when dry but can pucker when wet.

JLOTA COARCTATA

hairy calyptras;
leaves don't
curl; capsules
with exposed
stomates and
reflexed teeth;
base of leaf
with rows of
thick-walled
cells radiating
from the costa

F M

The **common *ulota*** of dry acid boulders and ledges, making small, dark tufts. Similar in the field to *Orthotrichum anomalum*, which though typically on limy rocks, sometimes grows with *hutchinsiae*. The best field characters for *hutchinsiae* are the densely hairy calyptras and capsules with 8 ribs and 8 reflexed teeth. The radiating cells at the leaf base are a good lab character.

ULOTA

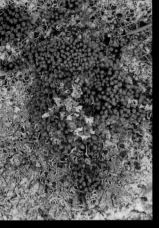

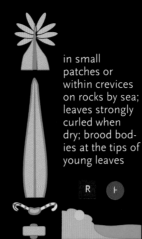

in small patches or within crevices on rocks by sea; leaves strongly curled when dry; brood bodies at the tips of young leaves

R F

A NORTHERN SPECIES, relatively large, strongly curly when dry, with elongate brood bodies on the tips of the youngest leaves. Forms patches on rocks by the sea, often within the spray zone but never submerged. Also reported from trees in coastal forests. *Schistidium maritimum* lives in the same habitat; it forms denser cushions and lacks brood bodies. Both seem to be rare. There are a lot of rocks on the northern coasts but very few have *Ulota* or *Schistidium*.

ULOTA PHYLLANTHA

WEISSIA CONTROVERSA C M Small, light yellow-green moss of disturbed soil; leaves slender, opaque, inrolled

plants light green, opaque; leaves slender, papillose, inrolled along the upper edges

A SMALL WEEDY MOSS that makes bright yellow-green patches in fields, borrow pits, river shores, and gravelly soil along roads and railroads. The leaves are slender, densely papillose, curled when dry, and strongly inrolled along their upper edges, wet or dry. It is found in all sorts of places but particularly common in open limy soil. *Barbula unguiculata* grows in the same sort of places and is the same color but lacks the inrolled edges. *Tortella inclinata*, a rare moss of limy habitats, has hooded tips that are somewhat similar. It grows in denser clumps and has shorter, broader leaves than *Weissia*.

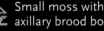

WEISSIA CONTROVERSA

ZYGODON CONOIDEUS R ⚲ Small moss with thick-walled multipapillose cells and axillary brood bodies, forming patches on tree bark

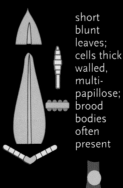

short blunt leaves; cells thick walled, multi-papillose; brood bodies often present

A SMALL RARE MOSS forming patches on tree trunks and, reportedly, concrete walls. Leaves short, blunt, flat edged, erect when dry; upper cells small, rounded, thick walled, multipapillose; club-shaped brood bodies often found in leaf axils. Almost all the northern forest records are from the north Atlantic coast. In the field it looks like a small *Orthotrichum* in patches rather than tufts.

ZYGODON CONOIDEUS

QUICK GUIDES TO HABITATS

QUICK GUIDES TO ACROCARPS

QUICK GUIDES TO PLEUROCARPS

QUICK GUIDES TO SPHAGNUM

ACROCARPS

PLEUROCARPS

SPHAGNUM

ABIETINELLA ABIETINA 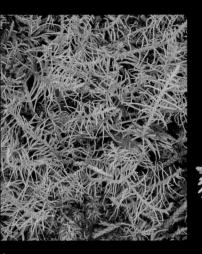 C M Large, wiry, light yellow-green, once-pinnate moss of open limy soil; branches taper to skinny tips

large, stiff, pinnate, in loose mats; paraphyllia (p) on stems; single papillae on leaf cells, several papillae on tip cell of branch leaves

A COMMON NORTHERN SPECIES of dry open limy soil and rocks, found from the temperate zone to the high arctic; common throughout the NFR. Like *Thuidium* but stringier and mostly once pinnate, often arching and making a loose thatch. *Haplocladium*, *Heterocladium*, and *Cyrto-hypnum* are similar and also grow on limy soil. In the field they are smaller, and their branches don't taper as conspicuously. Microscopically, *Heterocladium* has a double costa, *Cyrto-hypnum* has multiple papillae on the leaf cells, and *Haplocladium* has a single papilla on the tip cell of the branch leaves. *Rauiella,* also wiry and once-pinnate, is much smaller and grows on tree bark.

ABIETINELLA ABIETINA

AMBLYSTEGIUM  C Small stringy mosses with untoothed lanceolate leaves: cells short, alar cells not differentiated

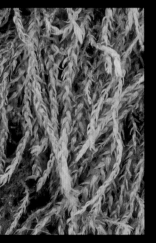

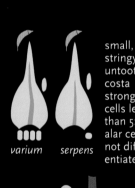

small, stringy, untoothed; costa strong; cells less than 5:1; alar cells not differentiated

varium   *serpens*

A COMMON GENUS of stringy creeping mosses with small sharp-pointed leaves and long, inclined capsules. In the field they don't look like much. *Homomallium*, *Platydictya*, *Platygyrium*, and *Pylaisia* have similar leaves; only the microscope really knows. *Varium* and *serpens* are common species on rocks, logs, and tree bases. They differ mostly in the length of their costas. *Hygroamblystegium*, typically on wet rocks in streams, differs in its thicker costa.

AMBLYSTEGIUM VARIUM   *AMBLYSTEGIUM SERPENS*

ANACAMPTODON SPLACHNOIDES F U Small moss with reflexed peristome teeth, making cushions in knotholes and crotches

small dense dark-green cushions on trees; oval costate leaves; capsules constricted below mouth, with reflexed teeth

AN EASTERN NORTH AMERICAN SPECIES, commoner south of the NFR, which makes small deep-green cushions in knotholes and crotches in trees. Leaves oval and pointed, their cells short; like *Amblystegium* but more erect and less of a creeper. Capsules are common and the best field character: they are erect and have a constricted neck and reflexed peristome teeth.

ANACAMPTODON SPLACHNOIDES

ANOMODON Stringy mat-forming papillose mosses with oval leaves, often blunt at the tip, on trees and limy rocks

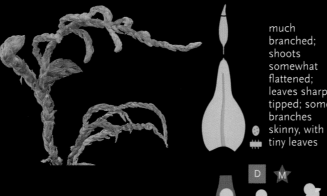

much branched; shoots somewhat flattened; leaves sharp tipped; some branches skinny, with tiny leaves

ANOMODON ATTENUATUS

COMMON MOSSES of tree bark, logs, and limy rocks, often stringy or shaggy, forming large patches from creeping stems. Leaves opaque, rounded at their bases, often narrowed to a parallel-sided tip.

Attenuatus, the commonest species, is medium sized and much branched. Some branches are blunt and club shaped; others end in skinny tips with tiny leaves. It makes thick mats on tree bases, logs, and moist limy rocks.

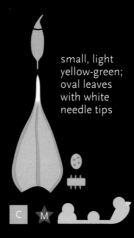

small, light yellow-green; oval leaves with white needle tips

ANOMODON ROSTRATUS

A SMALL DISTINCTIVE SPECIES with needle-tipped leaves that grows erect and makes light-colored mats on limy rocks. Commonly on boulders and ledges, often in moist places or in seepage. Identifiable on sight when in large thick mats, more puzzling when scattered among other species in crevices. The needle tips, combined with the tightly overlapping leaves, are the giveaway.

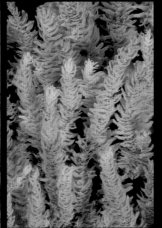

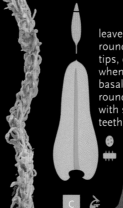

leaves with rounded tips, curled when dry; basal lobes rounded, with spiny teeth

ANOMODON RUGELII

THE NEXT THREE SPECIES are the coat-hook *Anomodons*, larger than *rostratus* and more erect and less branched than *attenuatus*. They are similar; we guess in the field and check in the lab.

Rugelii is a medium-sized tree-bark species, stringier and less common than *attenuatus*. The leaves have rounded clasping lobes at the base and a long, parallel-sided tip. Under the microscope, the lobes have spiny teeth. *Minor* is similar but with pointed basal lobes without spiny teeth.

QUICK GUIDES TO HABITATS

QUICK GUIDES TO ACROCARPS

QUICK GUIDES TO PLEUROCARPS

QUICK GUIDES TO SPHAGNUM

ACROCARPS

PLEUROCARPS

SPHAGNUM

broad blunt leaves with pointed decurrent lobes, not strongly curled when dry

OUR SECOND, and rarer, tree-bark *Anomodon,* also found on limy rocks. Medium sized, with somewhat flattened branches. Leaves with broad, rounded, parallel-sided tips and pointed basal lobes that run down the stem, not strongly twisted when dry. *Rugelii* is similar and can grow with *minor*; it is best distinguished by the rounded basal lobes with spiny teeth.

ANOMODON MINOR

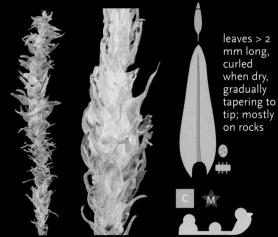

leaves > 2 mm long, curled when dry, gradually tapering to tip; mostly on rocks

THE COMMON LARGE *ANOMODON* of limy rocks, making large shaggy mats with the shoots curved up like fishhooks when dry. Only rarely on trees. Often grows with *attenuatus* and *rostratus*. Recognized by the large size and the leaves that taper from the base to a pointed tip, have pointed, decurrent basal lobes, and curl and arch when dry. *Minor,* which may grow with it on rocks, is half the size and has broader-tipped leaves that don't curl much. *Rugelii,* also smaller and mostly on trees, has rounded basal lobes with spiny teeth.

ANOMODON VITICULOSUS

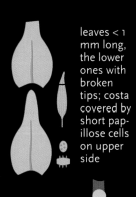

leaves < 1 mm long, the lower ones with broken tips; costa covered by short papillose cells on upper side

A SMALL OPAQUE MOSS making thin stringy mats on trees or, rarely, on rocks. Recognized in the field, when recognized at all, by the broken-off tips of the lower leaves. Confirmed in the lab by the papillose cells covering the costa on the upper side of the leaf.

BRACHYTHECIUM Light-green silky mosses in loose mats; leaves commonly pleated, with slender tips; costas slender; capsules short, dark, and curved

BRACHYTHECIUM CAMPESTRE*

large and weedy; pleated leaves with abruptly enlarged alar cells; smooth setae

COMMON AND OFTEN WEEDY MOSSES that grow on almost anything moist and a little dirty. Abundant in young woods, scarcer or absent in old growth and boreal forest. The genus is recognized, informally, by the loose, silky, light-colored plants and sharp-tipped leaves. The short capsules, single costa, and differentiated alar cells are a good confirmation. Recognizing the species is harder. Capsules help a lot. We guess in the field and check with a microscope.

Campestre (formerly *salebrosum*), found on wood and moist soil, is one of our commonest species. Its best marks are the pleated leaves, smooth setae, and large, inflated alar cells.*

BRACHYTHECIUM RUTABULUM

smooth leaves with alar cells in large decurrencies; papillose setae

LARGE AND COMMON, on wet wood and soil, often in swampy places. Resembles *campestre* but with larger, more inflated alar cells, nonpleated leaves, and papillose setae. The stem leaf tips are long acuminate. Can resemble, and be transitional to, *rivulare*, which has shorter and broader stem leaf tips.

BRACHYTHECIUM LAETUM

strongly pleated leaves with small teeth on margins and lots of square alar cells

A VERY COMMON WOODLAND MOSS, forming large patches on ledges, boulders, and tree bases. Locally dominant on limy rock. Leaves pleated, like *campestre*, but alar cells smaller, squarer, and more opaque. Capsules are rare; large sterile mats are common.

* *Campestre* is part of a difficult group of pleated-leaf species that includes *acuminatum, acutum, erythrorrhizon, falcatum, laetum,* and *rotaeanum*. The treatment of these species has changed greatly in the last few years. See Crum and Anderson (1981) for a traditional account and *Allen* (2014) for a recent one. And good luck.

QUICK GUIDES TO HABITATS

QUICK GUIDES TO ACROCARPS

QUICK GUIDES TO PLEUROCARPS

QUICK GUIDES TO SPHAGNUM

ACROCARPS

PLEUROCARPS

SPHAGNUM

small curly branches; leaves toothy, with short cells, small square alar cells, and the costa extending into tip; seta papillose

A COMMON SPECIES of logs, tree bases, and boulders, recognized in the field by the small size and curved branch tips. Confirmed by the short cells, leaf margins toothed to the base, costa extending into the leaf tip, small square alar cells, and papillose setae. *Brachythecium populeum,* which can grow with *reflexum,* has longer, more lanceolate leaves and longer cells.

BRACHYTHECIUM (SCIURO-HYPNUM) REFLEXUM

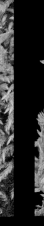

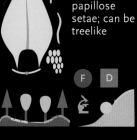

broad leaves with short tips; inflated alar cells in broad decurrencies; papillose setae; can be treelike

A LARGE SPECIES of wet places, often dominant on rocks in streams, on ledges with seepage, and in pools in swampy woods. Recognized in the field by concave leaves, spaced out along the stem, with short tips and large decurrencies. Confirmed by the bulging thin-walled alar cells and the short leaf tips. Some forms of *rutabulum* are similar and hard to place.

BRACHYTHECIUM RIVULARE

costa entering leaf tip; small square alar cells; setae papillose above

A COMMON WEEDY SPECIES of rocks and logs in open brushy woods and thickets. Light green and stringy but otherwise nondescript in the field; recognized under the microscope by the long costa, small square alar cells, and the weakly papillose seta. *Brachythecium campestre* looks similar in the field but has pleated leaves with shorter costas and larger alar cells.

BRACHYTHECIUM

small opaque alar cells; costa two-thirds of the leaf length or less; setae papillose above, smooth below

D

A MEDIUM-SIZED SPECIES that forms large shiny green mats on rocks in stream channels, often submerged at high water. The leaves may be curved to one side. The best technical characters are the small alar cells, the seta that is papillose below the capsule but not at the base, and the costa that stops below the leaf tip.

BRACHYTHECIUM (SCIURO-HYPNUM) PLUMOSUM

slender curved leaves with long points and teeth to base; small square alar cells; setae papillose to base

U

OUR SMALLEST *BRACHYTHECIUM*, forming glossy light-green mats on mineral soil. Sometimes recognizable from the slender curved leaves that make it look like a small *Hypnum*. Sometimes overlooked for the same reason. The best lab characters are the small alar cells, small teeth to base, and papillose seta.

BRACHYTHECIUM (BRACHYTHECIASTRUM) VELUTINUM

BROTHERELLA RECURVANS C

Glossy mat-forming moss of forest floors, with pinnate branching and curved leaves with inflated alar cells, sharp teeth, and no costas

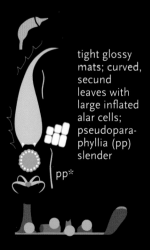

tight glossy mats; curved, secund leaves with large inflated alar cells; pseudoparaphyllia (pp) slender

pp*

A COMMON BUT HARD-TO-IDENTIFY MOSS of moist forest floors, making glossy green mats on tree bases, logs, and boulders. Looks much like a greener, tighter *Hypnum imponens*: pinnate branching, somewhat flattened shoots, leaves with the tips curved down. The sharply toothed margins and abruptly inflated alar cells are the key characters. We identify it presumptively by color and growth habit in the field but are not always right.

BROTHERELLA RECURVANS

* Pseudoparaphyllia are small appendages located at the base of the branches.

QUICK GUIDES TO HABITATS

QUICK GUIDES TO ACROCARPS

QUICK GUIDES TO PLEUROCARPS

QUICK GUIDES TO SPHAGNUM

ACROCARPS

PLEUROCARPS

SPHAGNUM

broad, decurrent stem leaves with twisted tips; tips of upper cells project

b s

C M

A LIGHT-GREEN MOSS with an arching, somewhat treelike growth form, common on wet ledges, seepy banks, and rocks by streams. Recognized by the broad costate stem leaves with twisted tips and decurrent bases. Confirmed by the minutely projecting upper cells and enlarged alar cells.

BRYHNIA NOVAE-ANGLIAE

slender leaves; sharp teeth on edges and costa; projecting upper cells; small alar cells

C

A SMALL MOSS, making cushions or thin mats on moist limy soil; often in recesses at the base of ledges or, as here, on banks where there is limy seepage. The sharp teeth and cells with projecting tips are distinctive under the microscope. In the field, we have to guess based on the habitat, light color, opacity, and sharp-tipped leaves.

BRYHNIA GRAMINICOLOR

BRYOANDERSONIA ILLECEBRA C M Large moss with thick cylindrical shoots and deeply concave oval leaves with twisted tips

thick cylindrical branches; deeply concave leaves with twisted tips

THE *GIANT WORM MOSS*, fat cylindrical shoots with closely overlapping, deeply concave leaves. The leaves are slightly wavy and have small, abruptly twisted tips. Common on seepy banks and forest floors in the Appalachians; largely restricted to the oak zone and to the southern edge of the northern forest.

CALLICLADIUM HALDANIANUM Yellow-green bushy moss with closely overlapping leaves on slightly flattened branches; leaves closely overlapping, without costas; long curved capsules with beaked lids

long arching capsules; slightly flattened branches; leaves with short costas, no teeth, and strongly inflated alar cells

A COMMON AND CONSPICUOUS MOSS of brushy thickets and second-growth woods, making large patches on tree bases and logs. Plants shaggy, irregularly branched, yellow green, with closely shingled leaves, flattened branches, and long curved or inclined capsules. Looks like *Entodon* but shaggier. One of the first mosses to colonize moist logs in young woods, and often one of the first to deceive young mossers. Good lab characters, for safety, are the long cells and strongly inflated alar cells.

CALLICLADIUM HALDANIANUM

CALLIERGON Large wetland mosses with broad, deeply concave leaves, strongly inflated alar cells and large clear cells (rhizoid initials) along the costa near the leaf tip

rhizoid initials near tip

concave hooded leaves; long costa; irregular branching; alar cells grade into upper cells and reach costa

LARGE MOSSES OF WETLANDS, often in pools that flood in the spring and dry up in the summer, with deeply concave costate leaves with rounded tips and strongly inflated alar cells. Easily confused with *Calliergonella* and *Scorpidium*, which lack costas, and *Pseudocalliergon* with wormlike branches and much smaller alar cells.

Cordifolium is a large common species, typically in pools in acid swamps, often growing with *Rhizomnium, Amblystegium, Pseudobryum,* and *Sphagnum.*

CALLIERGON CORDIFOLIUM

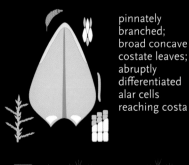

pinnately branched; broad concave costate leaves; abruptly differentiated alar cells reaching costa

THE COMMON *CALLIER-GON* of limy swamps and rich fens. Separated from *cordifolium* by the more regular branching and abruptly differentiated alar cells. This is one of several swamp mosses whose shoots die when submerged in the winter and then produce new growth when the water level drops in the spring.

CALLIERGON GIGANTEUM, with *Marchantia polymorpha*

QUICK GUIDES TO HABITATS

QUICK GUIDES TO ACROCARPS

QUICK GUIDES TO PLEUROCARPS

QUICK GUIDES TO SPHAGNUM

ACROCARPS

PLEUROCARPS

SPHAGNUM

sparsely branched; oblong leaves; inflated alar cells reach about half way to costa

CALLIERGON STRAMINEUM

A MUCH LESS STRIKING SPE-CIES than the two previous: slender, little branched, with oblong rather than oval leaves and inflated alar cells that do not reach the costa. Found in all sorts of wet open plac-es—ditches, pools, seeps, wet crevices in rocks—but rarely obvious or in large quantities. If you find a stringy, concave-leaved wetland moss with some rhizoids near the leaf tips, this is a good guess.

CALLIERGONELLA CUSPIDATA Big pinnate wetland moss with blunt concave leaves, inflated alar cells, and no costa

concave leaves with rounded tips, tapered branch tips, and no costa; large inflated alar cells

CALLIERGONELLA CUSPIDATA

A COMMON SPECIES of limy swamps and rich fens, re-sembling *Calliergon* and often growing with it. The tapering branch tips are a good field mark but not de-finitive. The ecostate leaves with rounded tips and a row of large alar cells are better. *Scorpidium*, which can grow with *Calliergonella*, has more pointed leaves with the tips often curved to one side.

CAMPYLIUM Large and small mosses with slender channeled needlelike leaf tips and differentiated alar cells*

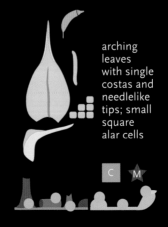

arching leaves with single costas and needlelike tips; small square alar cells

CAMPYLIUM (CAMPYLIADELPHUS) CHRYSOPHYLLUM

A GRAB-BAG GENUS, united by little more than the channeled leaf tips, subdivided in different ways by different authors. We have three wide-raging species, all found throughout the NFR. *Chrysophyllum* is the common upland species: medium sized, single costa, small square alar cells, and the standard *Campy-lium* leaf tip. It is found regularly on limy ledges and soil and, less regularly, on tree bases and logs. *Stellatum*, the common species of fens, has a short double costa.

* Split into three genera, *Campylium, Campyliadelphus,* and *Campylophyllum* in the *Flora of North America.* Nomenclator says that it's messy either way.

CAMPYLIUM

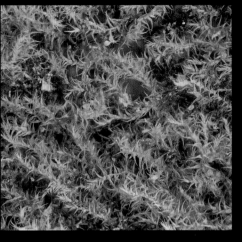

tiny, toothed leaves with arching tips and short double costas

A SMALL SPECIES of tree bark and rotting wood, recognized by the combination of slender arching leaf tips, short double costas, and teeth reaching the base of the leaves. Several other small rare species are similar and differ in microscopic details.*

CAMPYLIUM (CAMPYLOPHYLLUM) HISPIDULUM

arching leaves with long, channeled tips; short double costa; inflated alar cells

A LARGE WIDE-RANGING SPECIES of rich fens, often making continuous carpets over large areas. Found from the sub-tropics to the high arctic. The spreading leaves with straight needle tips are distinctive and easily seen in the field. *Limprichtia* and *Hamatocaulis*, which often grow with it, have slender leaves with curved tips. This is the only fen species with oval leaves with slender straight tips.

CAMPYLIUM STELLATUM

CIRRIPHYLLUM PILIFERUM   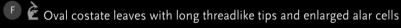 Oval costate leaves with long threadlike tips and enlarged alar cells

light green, small, pinnate; leaves abruptly narrowed to long threadlike tips

A WIDE-RANGING SPECIES of swampy shaded ground, found throughout the NFR but not common anywhere. Its distribution suggests that it may be a calciphile. Plants small, light green, loosely pinnate, with oval leaves with long, often twisted, threadlike tips. Nothing else in our flora has tips like this. *Bryhnia novae-angliae*, a common species that grows in the same sorts of places, looks generally similar but has much shorter leaf tips.

CIRRIPHYLLUM PILIFERUM

* *Campylium sommerfeltii* is a former variety of *hispidulum* promoted to species rank. *Campylium halleri* is a wide-ranging northern species that enters the NFR in eastern Canada.

QUICK GUIDES TO HABITATS

QUICK GUIDES TO ACROCARPS

QUICK GUIDES TO PLEUROCARPS

QUICK GUIDES TO SPHAGNUM

ACROCARPS

PLEUROCARPS

SPHAGNUM

CLIMACIUM D M Large treelike mosses with a central stem and spreading branches; stem leaves broad, branch leaves toothed

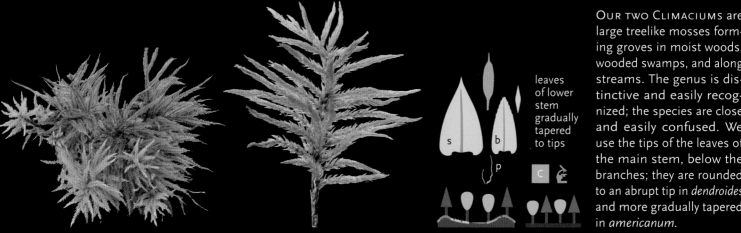

leaves of lower stem gradually tapered to tips

s b

p

C

OUR TWO CLIMACIUMS are large treelike mosses forming groves in moist woods, wooded swamps, and along streams. The genus is distinctive and easily recognized; the species are close and easily confused. We use the tips of the leaves of the main stem, below the branches; they are rounded to an abrupt tip in *dendroides* and more gradually tapered in *americanum*.

CLIMACIUM AMERICANUM

leaves of lower stem rounded to a short abrupt point

s b

C

CLIMACIUM DENDROIDES

DENDROIDES AND AMERICANUM overlap greatly in morphology, geography, and ecology. Both are common in the northern forest. *Dendroides* is overall the more northern and perhaps the one most likely to be encountered in the uplands; *americanum* ranges farther to the south and may be commoner in swamps. We see them, at best, at the weak end of what we can accept as species.

CRATONEURON FILICINUM F C Large dark messy moss of limy seepage and rich fens; broad, toothed stem leaves; short cells; a few scattered paraphyllia (p)

irregularly pinnate with hooked branch tips; stem leaves curved, with inflated alar cells, short upper cells, and slender tips

b s

p

A CHARACTERISTIC MOSS of limy seepage, found in springs, at the foot of banks, and where groundwater enters fens. Resembles several other genera. Good marks are the pinnate branching, curved tips to the branches, rhizoids along the stem, and broad-based stem leaves with slender, often curved tips. Under the microscope, the toothed leaves, short cells, large alar cells, and scattered paraphyllia are distinctive. *Palustriella falcata* is similar but has abundant paraphyllia and weakly differentiated alar cells.

CRATONEURON FILICINUM

CTENIDIUM MOLLUSCUM U 🔬 Leaves with long curved tips and short double costas; stem leaves broader than branch leaves*

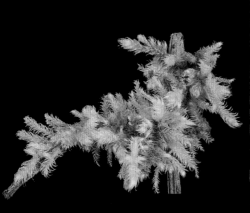

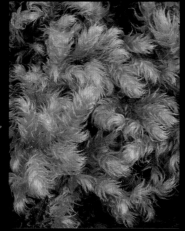

broadly oval stem leaves; falcate-secund branch leaves; tips of cells protruding; alar cells inflated

A WIDE-RANGING MOSS in both temperate and arctic zones; uncommon though locally plentiful in limy swamps in the NFR. Recognized by falcate-secund leaves, with short double costas, distinct teeth, and cells protruding at their tips. In the field it looks like a loose, fluffy, light-green *Hypnum* with out-of-control leaf tips. The broad stem leaves are a good field character.

CTENIDIUM MOLLUSCUM

CYRTO-HYPNUM U 🔬 Tiny threadlike mosses with regular pinnate branching and opaque multi-papillose leaves

tiny plants, once or twice pinnate, with oval leaves, multi-papillose cells, and unbranched paraphyllia (p); stems not papillose

TWO SPECIES OF SMALL MOSSES, related to *Rauiella* and *Thuidium* but considerably smaller. Stems wiry, opaque green, only 1–2 cm long, regularly once- or twice-pinnate; leaves 0.5 mm long or less, short tipped, with multi-papillose cells. Our species grow on soil, logs, or rocks, especially in limy areas. *Rauiella,* similar but generally bigger, has longer leaf tips, branched paraphyllia, and grows on trees. *Haplocladium* has single papillae. *Heterocladium,* which can look similar, has wide-spreading leaves with double costas and cells with protruding tips.

CYRTO-HYPNUM MINUTULUM

like *minutulum* but with papillose stems and branches and more often 2× pinnate

WE HAVE TWO SPECIES of *Cyrto-hypnum.* Both are widespread in temperate eastern North America but rare with us. *Minutulum,* with smooth branches, is found on soil and logs in moist forest floors. *Pygmaeum,* with papillose stems and branches, is more often twice pinnate and makes thin sparse mats on moist limy rocks in the shade.

CYRTO-HYPNUM PYGMAEUM

* Also called, informally but fondly, *Frizzy-whacky.*

QUICK GUIDES TO HABITATS

QUICK GUIDES TO ACROCARPS

QUICK GUIDES TO PLEUROCARPS

QUICK GUIDES TO SPHAGNUM

ACROCARPS

PLEUROCARPS

SPHAGNUM

slender, sharply folded leaves, over 1 mm wide; setae long, capsules exserted beyond perichaetial leaves.

F C

THREE SEMI-AQUATIC MOSSES, all wide spread and locally common in the NFR that grow on tree bases, dead wood and rocks in places that flood intermit tently. The leaves, which are arranged in three ranks, are long, slender, sharply folded, costate, and usually curved The alar cells are undifferentiated *Falcatum*, left, is a large species with broad leaves and exserted capsules *Capillaceum*, below, and *pallescens* (no shown) are smaller and have capsules surrounded by the perichaetial leaves *Capillaceum* has needle-tipped leaves *pallescens* blunter ones.

DICHELYMA FALCATUM

long slender leaves with costas excurrent into long needle tips; setae short, capsules surrounded by perichaetial leaves.

F C

A DISTINCTIVE SPECIES, found throughout eastern North America. Locally common in the NFR on seasonally flooded logs, rocks, shrubs, and tree bases by ponds and streams Immediately recognizable by the slender leaves with long needle tips formed by the ex current costas. Occurs in large masses, forming conspicuous black-green tufts on the stems of shrubs and large shaggy col lars on tree bases.

DICHELYMA CAPILLACEUM

leaves often curved; alar cells much enlarged; outer stem cells not enlarged; no rhizoid initials at leaf tip

AN AQUATIC MOSS of ponds and wetlands found throughout North America and common throughout the NFR Leaves slender, curved or straight, with large inflated alar cells ex tending from leaf edge to costa. *Warnstorfia* is similar but has rhizoida initials (enlarged clea cells) near the leaf tip.

DREPANOCLADUS ADUNCUS

* The genera *Hamatocaulis*, *Limprichtia*, *Sanionia*, and *Warnstorfia* have similar leaves and were formerly part of a broadly defined *Drepanocladus*. *Drepanocladus capillifolius*, not shown, is a northern species, very rare in the NFR, whose costa fills the leaf tip

ENTODON C M

Large shiny mosses with concave, closely overlapping leaves; alar cells small and square; upper cells short; capsules erect

branches strongly flattened, with distinct edges; 12 or fewer alar cells up margin

TWO COMMON SPECIES of tree bases, logs, and limy rocks, most often in second-growth deciduous woods. Recognized by the glossy gold-green color, closely overlapping oval leaves, and erect capsules. Our two species are close and often grow together. *Cladorrhizans* has strongly flattened branches and is a bit more of a northern species than *seductrix*.

ENTODON CLADORRHIZANS

branches round, without edges; 10 or more alar cells up margin

ENTODON SEDUCTRIX is a common species of logs and rocks, reaching its northern limit here. Similar to *cladorrhizans*, and like it glossy and yellow-green with tightly overlapping leaves. Differs only in the round or slightly compressed branches; the habitats seem to overlap completely.

ENTODON SEDUCTRIX

EURHYNCHIUM C F

Small mosses with toothy oval leaves, on moist soil; stem and branch leaves differ; upper cells short

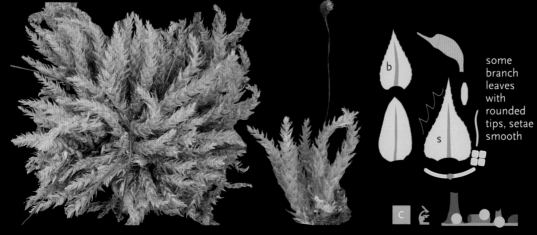

some branch leaves with rounded tips, setae smooth

TWO SPECIES OF SMALL MOSSES found on moist shaded soil, often with other mosses or among grasses and sedges. Recognized by the small, loosely spaced, short-pointed branch leaves contrasting with larger stem leaves. Under the microscope the leaves are toothed to the base and have single costas and short upper cells. *Pulchellum* is the commoner species in the northern forest, found on tree bases, logs, dirty rocks, and the forest floor. Its characters are branch leaves that vary from pointed to rounded and setae without papillae.

EURYNCHIUM (EURHYNCHIASTRUM) PULCHELLUM

QUICK GUIDES TO HABITATS

QUICK GUIDES TO ACROCARPS

QUICK GUIDES TO PLEUROCARPS

QUICK GUIDES TO SPHAGNUM

ACROCARPS

PLEUROCARPS

SPHAGNUM

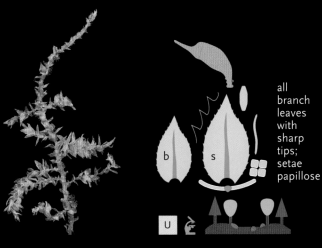

all branch leaves with sharp tips; setae papillose

A SIMILAR SPECIES, more southern and so less common in the northern forest. Found near streams or seeps, often in limy areas. Much like *pulchellum,* and like it a small creeping plant, easy to overlook. The best distinctions seem to be the uniformly sharp tips of the branch leaves and, when you can find one, the papillose seta.

EURYNCHIUM (OXYRRHYNCHIUM) HIANS, with *Plagiomnium ellipticum*

FONTINALIS F M   Limp long-stemmed aquatic mosses, often trailing in the current; large oval leaves without costas, arranged in three rows*

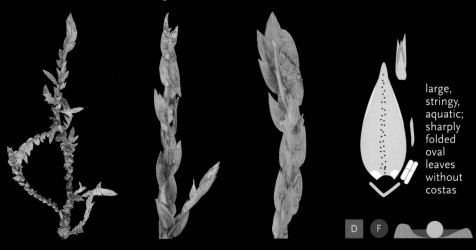

large, stringy, aquatic; sharply folded oval leaves without costas

A LARGE GENUS of aquatic mosses. Plants often large, limp, and stringy, dark green to black, attached to stones or wood, submerged for part or all of the year. Leaf shape varies with depth and current; submerged plants often differ from nearby ones on shores.

The common plants of northern streams have sharply folded leaves and are called *antipyretica.* Five other species without folded leaves are reported from the NFR but much less common. They are distinguished, unconvincingly, by small differences in leaf shape. We don't know them well enough to talk about them.

FONTINALIS ANTIPYRETICA

FORSSTROEMIA TRICHOMITRIA F R   Medium-sized pinnately branched moss making fringes on rocks and trees; leaves costate, with small square cells running up their margins

pinnately branched; oval leaves with a costa and a band of square cells running up margin

A SOUTHERN SPECIES of tree bark and limy rocks, rare with us. Shiny oval sharp-pointed leaves with a band of square cells up the margin; branches arch outwards, like coat hooks. Resembles *Leucodon* (which is only on trees) but has costate leaves and is more branched.

FORSSTROEMIA TRICHOMITRIA

* Bruce Allen, who monographed the group for the *Flora of North America*, warns "The taxonomy of *Fontinalis* is complicated by the small number of useful gametophytic characters." We concur and might add "and the low quality of the characters that there are."

HAMATOCAULIS VERNICOSUS Large pinnately branched moss with sickle leaves, often brown or red, in limy fens; leaves with a strong single costa; alar cells not enlarged; outer cells of stem thick walled

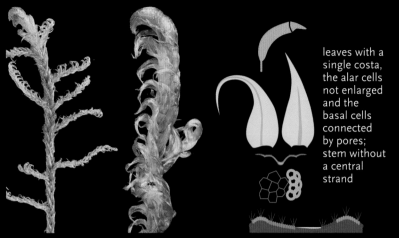

leaves with a single costa, the alar cells not enlarged and the basal cells connected by pores; stem without a central strand

A COMMON NORTHERN SPECIES of rich fens, with strongly hooked leaves. Resembles *Limprichtia*, *Warnstorfia*, and *Drepanocladus*, and can grow with them. Distinguished microscopically by negatives: alar cells not enlarged, stem cortex not inflated, central strand absent. Nomenclator says this makes him nervous.

HAMATOCAULIS VERNICOSUS

HAPLOCLADIUM VIRGINIANUM 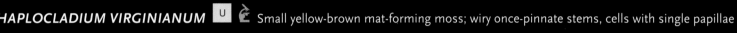 Small yellow-brown mat-forming moss; wiry once-pinnate stems, cells with single papillae

branch and stem leaves differentiated; cells with single papillae; single point on tip cell of branch leaves; paraphyllia (p) on stems

A SOUTHERN SPECIES of soil, rocks, and tree bases, usually in the open. Uncommon in the northern forest. Plants yellowish, opaque, wiry, once-pinnate, resembling a small, less branched, *Thuidium*. Several similar genera are best distinguished microscopically. *Haplocladium* has single papillae and single point on the tip cell of the branch leaves. *Abietinella* is taller, often arches, and has strongly tapered branches and multiple points at the branch leaf tips. *Rauiella* and *Cyrto-hypnum* have multi-papillose cells. *Heterocladium* has a short double costa and is usually in boreal forest.

HAPLOCLADIUM VIRGINIANUM

HELODIUM Once-pinnate mosses with broadly oval stem leaves and paraphyllia on the stems and leaf bases

large, regularly pinnate; oval stem leaves with short points; abundant paraphyllia (p) on stems and leaf bases; tips of cells protrude

A CHARACTERISTIC NORTHERN MOSS of limy fens and seeps, often growing amidst sedges and *Sphagnum* in hummocks. Large, bright yellow-green, stiffly pinnate, with broadly oval or heart-shaped stem leaves with paraphyllia at their bases. Found throughout the NFR, and common locally in the proper habitats; often with *Tomentypnum*, *Calliergonella*, *Scorpidium*, *Sphagnum warnstorfii,* and other rich-fen species.

*HELODIUM BLANDOWII**

* Written *Elodium* in the *Flora of North America*. In 1870 Sullivan transcribed the Greek ῾ελοδεσ without an initial *h*. In 1905, Warnstorf supplied one. The nomenclatural code is vague, their descendents still battle.

QUICK GUIDES TO HABITATS

QUICK GUIDES TO ACROCARPS

QUICK GUIDES TO PLEUROCARPS

QUICK GUIDES TO SPHAGNUM

ACROCARPS

PLEUROCARPS

SPHAGNUM

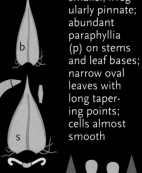

smaller, irregularly pinnate; abundant paraphyllia (p) on stems and leaf bases; narrow oval leaves with long tapering points; cells almost smooth

A SMALLER SPECIES of wooded swamps, local and uncommon in the southern parts of the NFR. The best marks are the narrow-oval stem leaves with long acuminate points and the abundant paraphyllia on the stem. Differs from *blandowii* in its smaller size, less regularly pinnate branching, and cells that are smooth or only weakly papillose.

HELODIUM PALUDOSUM

HERZOGIELLA

Small mosses of shaded soil or moist wood with wide-spreading leaves with slender tips, teeth almost to their bases, and outer stem cells inflated. They fruit abundantly in early spring.

slender wide-spreading leaves with slender tips, toothed to the base; short costas; inflated alar cells that run down the stem; inflated stem cortex

A SMALL COMMON MOSS that makes large mats on shaded soil banks and dirty ledges. Recognized in the field by the wide-spreading, loosely arranged leaves with slender tips and the capsules that appear early in the spring; confirmed by the strong teeth, weak costa, inflated alar cells that run down the stem, and inflated stem cells.

HERZOGIELLA STRIATELLA

shoots flattened; leaves widely spaced, toothed to base, with just a few enlarged alar cells that don't run down the stem; fruits abundantly in spring

A SMALL NONDESCRIPT MOSS, uncommon or overlooked, found on rotting wood in moist forests. Shoots flattened, leaves widely spreading, costa short, leaves toothed to the base; alar cells slightly enlarged, not running down stem. Resembles *Plagiothecium*, *Pseudotaxiphyllum*, and *Taxiphyllum*: confirm it under the scope.

HERZOGIELLA TURFACEA

HETEROCLADIUM DIMORPHUM Once-pinnate moss of limy soil, with a short double costa, wide-spreading leaves, and long slender points on the stem leaves

heart-shaped stem leaves with long slender tips and double costas, papillose by protruding cell tips

A SMALL RARE NORTH-ERN MOSS of limy soil and ledges. Recognized in the field by the once-pinnate shoots, leaves that spread widely when wet, and long curled tips of the stem leaves when dry. Confirmed microscopically by its double costa bordered by elongate smooth cells and the short outer and upper cells with protruding cell tips.

HETEROCLADIUM DIMORPHUM

HOMALIA TRICHOMANOIDES Distinctive shiny golden moss with flattened shoots and spoon-shaped leaves

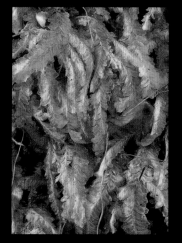

oboval finely toothed leaves with a single costa, in flattened shoots

AN UNUSUAL MOSS, resembling a leafy liverwort in its flatness, though costate and shinier than any liverwort. Common, though often inconspicuous, on trees in fertile woods and on moist, shaded, ledges. Look for small glossy yellow plants mixed in with other mosses.

HOMALIA TRICHOMANOIDES

HOMOMALLIUM ADNATUM Small shiny moss making traceries on rocks; spoon-shaped leaves with short costas and small square alar cells

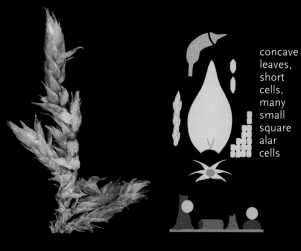

concave leaves, short cells, many small square alar cells

ONE OF OUR COMMONEST ROCK MOSSES, making thin, shiny, irregularly branched mats. Most common on limy rocks but also on acid rocks and trees. Recognized in the field by the shininess and the concave leaves with short tips. Confirmed by the short costas and short cells. *Hypnum pallescens* is more regularly branched and has teeth; *Amblystegium* has flatter, pointier leaves with strong costas; *Pterigynandrum* is dark and dull; *Platydictya* is tiny.

HOMOMALLIUM ADNATUM

QUICK GUIDES TO HABITATS

QUICK GUIDES TO ACROCARPS

QUICK GUIDES TO PLEUROCARPS

QUICK GUIDES TO SPHAGNUM

ACROCARPS

PLEUROCARPS

SPHAGNUM

HYGROAMBLYSTEGIUM Small irregularly branched mosses on rocks in streams; leaves oval, with short cells and a broad costa

sharp pointed tips; short cells; costa about 0.05 mm wide at base

COMMON MOSSES, stringy and dark green, with pointed leaves, short tips, and undifferentiated alar cells. Typically found on rocks in and near streams; also on wet ledges. Differing from *Amblystegium* only in the wetter habit and the broader costas. *Tenax* is our common species; *fluviatile* is similar but has an even broader costa and a blunter tip. Among the Amblystegiums, a little bit of morphology goes a long way.

HYGROAMBLYSTEGIUM TENAX

HYGROHYPNUM Among the commonest mosses of rocks in stream channels; concave leaves with short or blunt tips that are often hooked at the branch tips; long cells; extremely variable costas; alar cells often differentiated

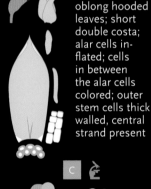

oblong hooded leaves; short double costa; alar cells inflated; cells in between the alar cells colored; outer stem cells thick walled, central strand present

A COMMON GENUS, found in rocky stream channels and on wet ledges. Recognized by concave leaves with long cells and short tips that tend to curve in one direction near the branch tips. The genus is identifiable in the field; the species need the microscope. *Eugyrium*, with a short double costa, yellow basal cells, and abruptly inflated alar cells, is one of our commonest species. *Sematophyllum marylandicum* is very similar but has flatter leaves without hooded tips, no colored basal cells, and stems without a central strand.

HYGROHYPNUM EUGYRIUM

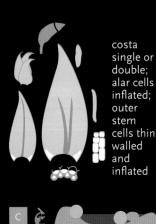

costa single or double; alar cells inflated; outer stem cells thin walled and inflated

A SECOND COMMON SPECIES, resembling *eugyrium* but with more pointed leaves and thin-walled stem cells that strip off with the leaves. The alar cells are strongly inflated; the costas may be single or double. Common on rocks in streams, often submerged; also in cascades and on ledges with seepage.

HYGROHYPNUM

HYGROHYPNUM DURIUSCULUM*

leaves broadly oval; costa single or double; a few alar cells strongly enlarged

F U

A MEDIUM-SIZED SPECIES with blunt, broadly rounded leaves and a few enlarged cells at the extreme corners of the leaf. Costas mostly short and double, occasionally single. Found in brooks and cascades and sometimes on limy rocks. The broad leaves are a good field character, the few enlarged alar cells a good confirmation. The rare *H. bestii,* not shown, is similar but has long cells along the margins.

HYGROHYPNUM SUBEUGYRIUM

dark except at tips; some tips rounded and toothed; alar cells small and square, not inflated or only the outermost inflated; no central strand

R

A RARE SPECIES of rocks in brooks, known at scattered stations from Newfoundland and Nova Scotia to New York, with a disjunct record in Minnesota. Small plants with the young shoots green and the older ones red brown; leaves with short double costas, falcate secund near the branch tips. Close to *eugyrium* but with broader apices that are often finely toothed, small square alar cells, and stems with a thick cortex and no central strand. Bruce Allen says it is common in Maine; it may be overlooked elsewhere.

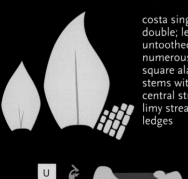

costa single or double; leaves untoothed, with numerous small square alar cells; stems with a central strand; limy streams and ledges

U

HYGROHYPNUM LURIDUM

very small; costa usually short and double; leaves with tiny teeth to the base; few square alar cells; stems without a central strand; mountain ledges and streams

R

HYGROHYPNUM MONTANUM

TWO UNCOMMON SPECIES with leaves under 2 mm and usually somewhat falcate secund. *Luridum,* wide ranging in northern North America, has a variable costa, untoothed leaves, and small square alar cells. It is occasional on limy rocks in streams and on wet limy ledges in the NFR. *Montanum,* our smallest *Hygrohypnum,* is tiny and rare. Its marks are the small leaves, mostly under 1 mm long, with fine teeth near the tip, a short double costa, and a few small square alar cells. There are NFR records are from Vermont, New Hampshire, and northern New York. We have looked for it but not found it.

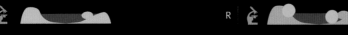

* Previously called *Hygrohypnum molle.* That name is now restricted to a species of western North America.

QUICK GUIDES TO HABITATS

QUICK GUIDES TO ACROCARPS

QUICK GUIDES TO PLEUROCARPS

QUICK GUIDES TO SPHAGNUM

ACROCARPS

PLEUROCARPS

SPHAGNUM

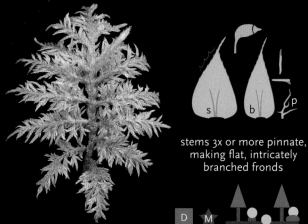

stems 3x or more pinnate, making flat, intricately branched fronds

D M

LARGE MOSSES with arching fronds, growing over rocks, logs, and soil in forests, often in large patches. Stem leaves often wider than branch leaves. *Splendens* is our most common species, recognized by the 3× or 4× pinnate fronds and stepladder growth, each frond arching above the one from the previous year. It is a common boreal forest dominant, growing on anything near the ground, often in or over other mosses. Differs from other species of the genus in its more pinnate fronds and cells that protrude at the tips.

HYLOCOMIUM SPLENDENS

twice-pinnate; stem leaves with long tips, double costas, and small teeth

F U

A LARGE SPECIES of moist shaded boulders and ledges, often with limy seepage, making big thick shaggy mats. Mostly Appalachian, local in the eastern NFR, absent from the western NFR. The fronds are twice-pinnate; the stem leaves are rounded at the base and pinched in just below the long skinny tip.

HYLOCOMIUM (LOESKEOBRYUM) BREVIROSTRE

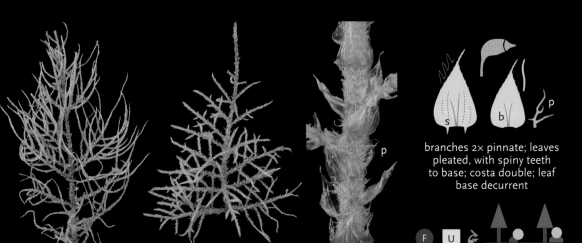

branches 2× pinnate; leaves pleated, with spiny teeth to base; costa double; leaf base decurrent

F U

AN UNCOMMON SPECIES of rocks and logs in moist conifer forests and swamps. The plants look like a stringy, less pinnate version of *Hylocomium splendens* that hasn't filled out. Confirmed by the pleated, spiny-toothed leaves, mostly under 2 mm long, some with double costas. The species is wide ranging; in the NFR we see it most commonly in the mountains and the north.

HYLOCOMIUM (HYLOCOMIASTRUM) UMBRATUM

* The four North American species of *Hylocomium* are split into three genera (*Hylocomium*, *Hylocomiastrum*, *Loeskeobryum*) in the *Flora of North America* and elsewhere. The differences between the genera are small and not very consistent. They have many characters in common, and one genus suits us fine.

HYLOCOMIUM

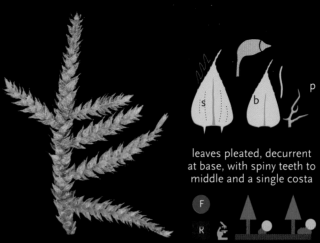

leaves pleated, decurrent at base, with spiny teeth to middle and a single costa

A NORTHERN SPECIES, rare with us, found on the boulders and on the forest floor in moist conifer woods. Recognized by the pleated leaves with spiny teeth and a single costa. *Umbratum* is similar but has leaves with double costas that are toothed to the base. *Pleurozium schreberi*, surprisingly similar, has shorter leaf tips and short double costas.

HYLOCOMIUM (HYLOCOMIASTRUM) PYRENAICUM

HYPNUM C M

Pinnately branched mosses with slender ecostate sickle-shaped leaves; tips curved towards the substrate; alar cells differentiated; pseudoparaphyllia (pp) usually present and useful for identification

some alar cells thick walled and yellow; leaves falcate secund, weakly toothed; pp. broad and lobed

VERY COMMON MOSSES of forests and wetlands with falcate-secund leaves whose tips curl down, making the shoots look braided. *Ptilium*, *Ctenidium*, and *Brotherella* are similar. *Ptilium* has erect fernlike fronds, *Ctenidium* has broad stem leaves and protruding cells, *Brotherella* has long wispy tips and abruptly inflated alar cells.*

H. imponens is our commonest species, found on rocks, logs, and tree bases everywhere, making continuous carpets in boreal forests. It is identified by the large size, pinnate branching, and yellow-green color in the field. It is confirmed by the colored alar cells and incised pseudoparaphyllia (pp).

HYPNUM IMPONENS

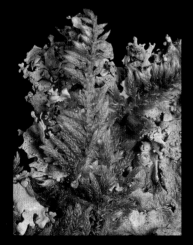

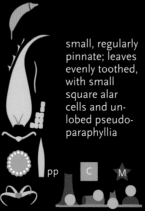

small, regularly pinnate; leaves evenly toothed, with small square alar cells and unlobed pseudoparaphyllia

A SMALL COMMON MOSS of rocks, tree bases, and logs, like a half-size *imponens*. Recognized in the field by the slender shoots, evenly pinnate branching and nice braided look. Confirmed by the toothed leaf tips, small alar cells, and unlobed pseudoparaphyllia (pp). Inconspicuous but pretty, and easy to find if you look.

HYPNUM PALLESCENS

* Several rare Hypnums (*curvifolium, fertile, pratense, plicatulum*) occur in the NFR. We see them too rarely to present them here.

133

QUICK GUIDES TO HABITATS

QUICK GUIDES TO ACROCARPS

QUICK GUIDES TO PLEUROCARPS

QUICK GUIDES TO SPHAGNUM

ACROCARPS

PLEUROCARPS

SPHAGNUM

irregular branching; inflated alar and outer stem cells; no pseudoparaphyllia

THE COMMON WETLAND HYPNUM, making mats or cushions in pools and on hummocks, often mixed with grasses and sedges. A large irregularly branched species, best identified by the inflated alar cells and outer stem cells. Species of *Drepanocladus*, *Warnstorfia*, and *Sanionia* can look similar; they are costate, but it is difficult to tell this in the field.

*HYPNUM LINDBERGII**

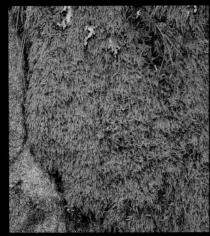

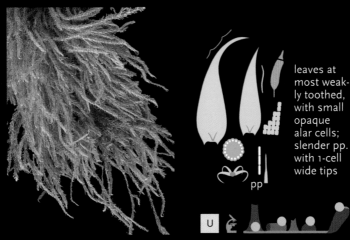

leaves at most weakly toothed, with small opaque alar cells; slender pp. with 1-cell wide tips

A SMALL WIDE-RANGING SPECIES found on trees, logs, and rocks in both temperate and boreal forests. Often slender, elongate, irregularly branched, or dangling. Ecologically tolerant: found in dry and wet places, and on acid and limy rocks. Hard to identify in the field, both because it is variable and because *H. imponens* can also dangle. The best characters are the numerous small alar cells and the unlobed pseudoparaphyllia (pp) with 1-cell wide tips.

HYPNUM CUPRESSIFORME

ISOPTERYGIOPSIS MUELLERIANA Ⓕ Ⓤ 🔬 Small flattened moss; leaves slender-tipped, shoots tapering to the tip and base

short double costa; flattened shoots; outer stem cells inflated, alar cells not differentiated; leaves not decurrent

A SMALL SHINY FLATTENED MOSS with narrow tapering shoots, inflated outer stem cells, and undifferentiated alar cells. Usually found in humus on or near moist limy ledges. Resembles, in a general way, *Plagiothecium* and its segregates *Pseudotaxiphyllum* and *Taxiphyllum*. Can be recognized in the field by the narrow featherlike shoots but needs to be confirmed in the lab. Sue says "It isn't much but it is something."

ISOPTERYGIOPSIS MUELLERIANA

* Bruce Allen, in *Maine Mosses*, makes an interesting argument for transferring this species to *Calliergonella* as *C. lindbergii*. Nomenclator says he is planning a new-genus party for it and you all are invited.

ISOTHECIUM MYOSUROIDES Silky, light yellow-green, irregularly frondose moss that forms fringes on boulders and ledges in cold woods; leaves sharply toothed, with small square alar cells

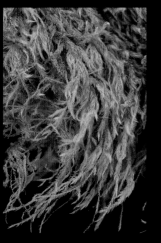   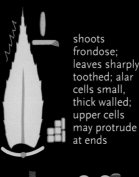

shoots frondose; leaves sharply toothed; alar cells small, thick walled; upper cells may protrude at ends

A WIDESPREAD MOSS of western North America, uncommon and restricted to the northern Atlantic Coast with us. Typically found as fringes on boulders in cold conifer or mixed woods. Recognized in the field by the yellow-green color, frondose shoots, and curved branches. Under the scope, the sharp teeth, single costa, and thick-walled alar cells are diagnostic. *Hypnum cupressiforme* and *imponens* are often abundant in the same woods and can look quite similar.

ISOTHECIUM MYOSUROIDES

LEPTODICTYUM RIPARIUM 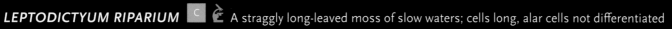 A straggly long-leaved moss of slow waters; cells long, alar cells not differentiated

long costate leaves with long cells; alar cells not differentiated; long, inclined capsules

A COMMON MOSS of slow waters, found in ponds, marshy streams, and pools in swamps, often on submerged wood. Differing from *Amblystegium* in its longer skinnier leaves with long cells; and from *Drepanocladus* in the small alar cells. Always check it with the microscope; aquatic mosses are super variable.

LEPTODICTYUM RIPARIUM

LESKEA Small stringy light yellow-green mosses with oval leaves, a single costa, and small papillose cells

  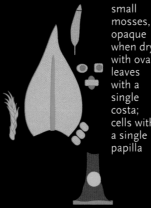

small mosses, opaque when dry, with oval leaves with a single costa; cells with a single papilla

COMMON MOSSES OF TREE BARK and recently fallen logs, making dense, light-green mats. Leaves small, oval, short pointed, and papillose; alar cells small and square, capsules erect. *Leskea, Platygyrium,* and *Pylaisia* are our three common tree-bark creepers. Only *Leskea* is light green and papillose.

We have three species, *polycarpa, obscura,* and *gracilescens,* separated by small differences in leaf shape, often obscured by variation. We only name them when we have to, and then with our fingers crossed.

LESKEA POLYCARPA GROUP

QUICK GUIDES TO HABITATS

QUICK GUIDES TO ACROCARPS

QUICK GUIDES TO PLEUROCARPS

QUICK GUIDES TO SPHAGNUM

ACROCARPS

PLEUROCARPS

SPHAGNUM

135

LESKEELLA NERVOSA Small northern moss of tree bark making dark thin mats; costa fills leaf tip; brood branches (bb) often present

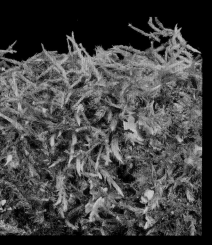

slender leaves with the costa filling the tip; small oval cells; clusters of brood branches near the branch tips

bb

A SMALL SCRUFFY DARK-GREEN MOSS making thin mats on tree bark and rocks. Recognized by the leaf tip filled with the costa and the clusters of brood branches at the branch tips. Under the microscope the small smooth oval cells are distinctive. *Platygyrium*, our commonest tree-bark creeper, also has brood branches and slender leaf tips. It is a shinier moss, with longer cells and no costa.

LESKEELLA NERVOSA

LEUCODON ANDREWSIANUS M Large moss making shaggy fringes on large trees; branches cylindrical, curved outward; leaf tips point to one side; brood branches common

cylindrical coat-hook branches; sharp-pointed oval leaves with a band of square cells up the edge; no costa

THE *COAT-HOOK* MOSS, common on the trunks of old trees in old woods, with cylindrical sparsely branched shoots arching out from the trunk and making curly fringes. Often with *Neckera pennata* and the liverwort *Porella platyphylla*, both of which have strongly flattened branches with the leaves in two rows. *Forsstroemia trichomitria*, a rare moss, is similar to *Leucodon* and may occur on trees. It is pinnately branched and has costate leaves.

LEUCODON ANDREWSIANUS

LIMPRICHTIA REVOLVENS U Large golden or brown sickle-leaved moss of rich fens; alar and outer stems cells inflated

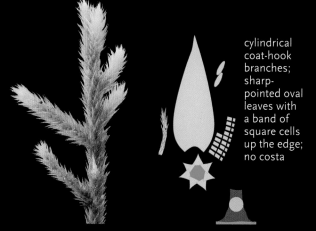

long curved leaf tips; inflated alar and outer stem cells; stems with a central strand

A DISTINCTIVE NORTHERN MOSS of rich fens and limy seepage, uncommon and scattered in the NFR. Suggested in the field by the habitat and deep red-gold color. Confirmed by the inflated alar and cortical cells. *Hamatocaulis*, which can be reddish too, has no inflated cells. *Limprichtia cossonii*, rare and northern, is said to have thinner-walled basal cells. We have not seen it.

LIMPRICHTIA REVOLVENS

LINDBERGIA BRACHYPTERA Small uncommon moss of tree bark, with needle tips, papillose cells and leaves wide spreading when wet

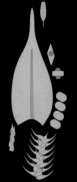

dry leaves with white needle tips; wet leaves spring open; cells papillose

SMALL DARK STRINGY MOSS of tree bark, either uncommon or overlooked. Found throughout the eastern United States but scarcer northwards. Recognizable, once you have found it, by the whitish needle tips of the dry leaves and the way the leaves spring open when you wet them. It is not showy, and finding it to recognize is the problem.

LINDBERGIA BRACHYPTERA

MYURELLA C M Small papillose mosses with broadly oval, concave leaves; always on limy rocks or in limy seepage

small spoon-shaped leaves, only loosely over-lapping, with needle points, spiny teeth, and strongly papillose cells

SMALL DISTINCTIVE MOSSES found in recesses in limy rocks and on soil with limy seepage; recognized by their small size, broad-oval leaves, and light white-green color.

The species are northern but not restricted to the mountains. *Sibirica*, with separated leaves with spiny teeth and high papillae, occurs almost anywhere there are sheltered, seepy crevices in limy rocks. In cool places it may form extensive mats; in warm ones there may be only a few shoots, mixed with other mosses.

MYURELLA SIBIRICA

small, tightly over-lapping, spoon-shaped leaves with short points and low teeth, making cylindrical shoots; upper cells protrude slightly at their tips

A SMALL NORTHERN SPECIES with wormlike cylindrical shoots, found on cliffs and in swamps, always in limy seepage. The erect stems originate from horizontal creeping ones and can form dense mats. It is most commonly found in crevices in cold limy cliffs, but we have also seen it in limy seepage in swamps and deciduous woods. *Anomobryum* and *Plagiobryum* also have cylindrical shoots and can grow with *Myurella*. They are acrocarps with smooth cells that do not form dense mats.

MYURELLA JULACEA

QUICK GUIDES TO HABITATS

QUICK GUIDES TO ACROCARPS

QUICK GUIDES TO PLEUROCARPS

QUICK GUIDES TO SPHAGNUM

ACROCARPS

PLEUROCARPS

SPHAGNUM

NECKERA PENNATA C M Large shiny moss making curled fringes on trees; flattened shoots, ripply leaves

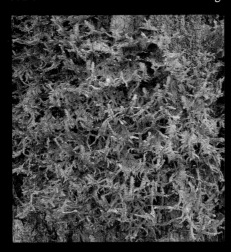

NECKERA PENNATA

leaves oval, wavy, often in two rows, the lower edges often inrolled

A CHARACTERISTIC MOSS OF OLD TREES and old forests, found high off the ground, recognized by the flat shoots and ripply leaves. Often associated with *Leucodon*, the liverwort *Porella*, and the lichen *Lobaria*, which make fringes as well. Very common on old maples and ashes with large crowns. *Neckera complanata* (not shown), which occurs rarely in the Maine and the Maritimes, is smaller than *pennata* and has blunter leaves that aren't wavy.

PALUSTRIELLA FALCATA R A rare moss of limy northern cliffs and fens, with slender sickle-curved leaves, a single costa, and slender paraphyllia on the stem.

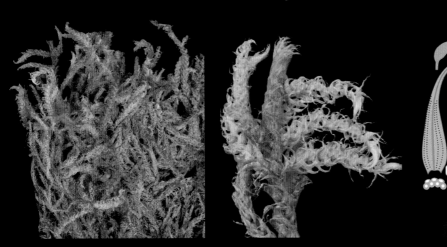

PALUSTRIELLA FALCATA*

costate, slender, strongly curved leaves with pleats; basal cells thick walled; paraphyllia on stems

R

AN UNCOMMON NORTHERN SPECIES, widespread in western North America, rare with us and mostly found on cold wet limy cliffs. Basically a *Drepanocladus* with paraphyllia (p) on the stems. Medium sized, often yellow brown, with slender, long-tipped, costate leaves and slender paraphyllia. Basal cells thickened; stem cortex not inflated; alar cells not inflated. Probably indistinguishable in the field from *Sanionia*, *Hamato-caulis*, *Drepanocladus*, and others. The paraphyllia are the key.

PLAGIOTHECIUM D Large or small mat-forming mosses with sparsely branched shoots from creeping stems; untoothed decur-rent leaves with short costas; outer stem cells inflated; elongate brood bodies on leaves or in leaf axils.

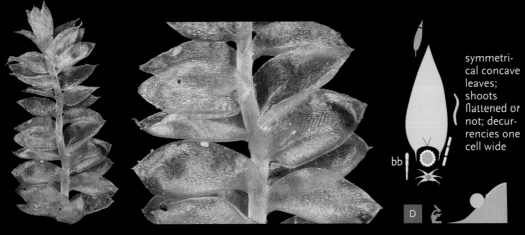

PLAGIOTHECIUM CAVIFOLIUM

symmetri-cal concave leaves; shoots flattened or not; decur-rencies one cell wide

bb

D

COMMON MOSSES of wet dirty ledges, also found on logs, tree bases, and boulders, growing from colorless creeping stems which produce the leafy shoots. Charac-terized, weakly, by the decurrent leaves and inflated stem cortical cells. Similar to *Isopterygiopsis*, *Pseudotaxiphyllum*, and *Taxiphyllum* and hard to tell from these without a microscope.

Cavifolium is a large common spe-cies that makes big mats on ledges and can cover large areas. Its marks are its symmetrical concave leaves, weakly or not at all flattened shoots, and slender decurrencies.

* Formerly called *Cratoneuron commutatum*.

PLAGIOTHECIUM

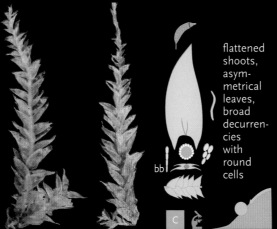

flattened shoots, asymmetrical leaves, broad decurrencies with round cells

bb

C

PLAGIOTHECIUM DENTICULATUM

A COMMON SPECIES of wet rocks, making fringes on dripping ledges and in small channels. Mature shoots strongly flattened, decurrencies broad, with round cells. Quite similar to *laetum* in the field; distinguished microscopically by the larger cells and broader decurrencies. A strongly flattened *Plagiothecium* is one of these two, unless of course it is not a *Plagiothecium* at all. A *Plagiothecium* with nonflattened or weakly flattened shoots is likely *cavifolium*, though with the same proviso.

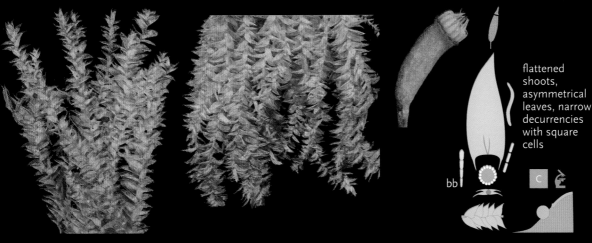

flattened shoots, asymmetrical leaves, narrow decurrencies with square cells

bb

C

PLAGIOTHECIUM LAETUM

THE THIRD OF OUR COMMON PLAGIOTHECIUMS, smaller than the previous two and occurring on tree bases and logs as well as wet cliffs. The flattened shoots and asymmetrical leaves are like *denticulatum*, from which it differs in the smaller cells and narrower decurrencies. Field identification is risky.

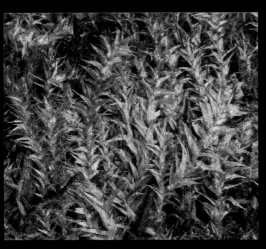
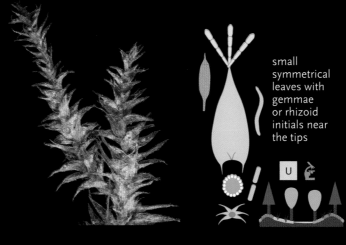

small symmetrical leaves with gemmae or rhizoid initials near the tips

U

PLAGIOTHECIUM LATEBRICOLA

A TINY, NORTHERN SPECIES, found on rotting wood near streams and in swamps. Leaves with enlarged cells near the tips which produce elongate gemmae or rhizoids. Can be recognized in the field if the gemmae are present; our other species produce brood branches or gemmae at the bases of the leaves but not at the leaf tips.

QUICK GUIDES TO HABITATS

QUICK GUIDES TO ACROCARPS

QUICK GUIDES TO PLEUROCARPS

QUICK GUIDES TO SPHAGNUM

ACROCARPS

PLEUROCARPS

SPHAGNUM

PLATYDICTYA [F] [U] Tiny, sparsely branched mosses making thin mats; leaves under 0.5 mm long, without costas

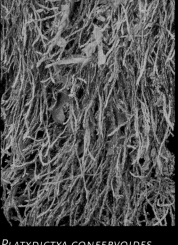

leaves untoothed; alar cells barely differentiated; branch angles wide

[U]

PLATYDICTYA CONFERVOIDES

OUR SMALLEST PLEUROCARPS: tiny, stringy mosses with leaves 0.2 to 0.5 mm long and barely visible with a lens. Costa absent or very weak, upper cells short. The two species shown here make thin mats on limy rocks. A third, *subtilis,* is a rare species of tree bark. All are uncommon. The genus can be recognized by the small size; a microscope is needed for the species.

Confervoides occurs regularly on dry limy rocks. It has wide branch angles, untoothed leaves, and largely undifferentiated alar cells.

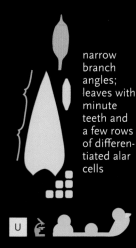

narrow branch angles; leaves with minute teeth and a few rows of differentiated alar cells

[U]

A TINY NORTHERN SPECIES, found on limy rocks and soil. It ranges high into the arctic and is uncommon here. The best marks are the narrow branch angles, more differentiated alar cells and minute teeth. It is also said to have smooth rhizoids.

PLATYDICTYA JUNGERMANNIOIDES, with *Homomallium adnatum* and *Brachythecium laetum*

PLATYGYRIUM REPENS [C] [M] Small shiny moss of tree bark: slender pointed leaves; erect branches with brood branches (bb)

brood bodies at thickened tip

leaves with long slender points, recurved edges, and small square alar cells; erect shoots with brood branches at their tips

bb

THE COMMONEST CREEPING MOSS ON TREE BARK: small, shiny yellow green or dark green, sharp leaved, ecostate, making flat mats. Brood branches confirm the identification. Plants without brood branches can look much like *Pylaisia,* also common on trees. The straighter branches and the recurved edges of the leaves of *Platygyrium* often, but not always, separate them.

PLATYLOMELLA LESCURII U 🔬 Small dark stringy moss with oval bordered leaves, found on rocks in streams

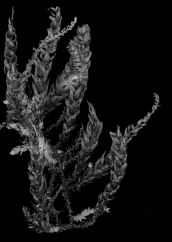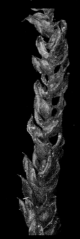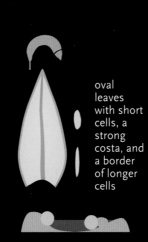

oval leaves with short cells, a strong costa, and a border of longer cells

A SMALL COSTATE MOSS with oval, bordered leaves, found on rocks in streams, usually submerged at high water. Resembles *Hygroamblystegium tenax,* which lacks borders. The border is diagnostic but hard to see in the field. An Appalachian species, common south of us, scattered and uncommon in the eastern parts of the NFR, absent from the western Great Lakes.

PLATYLOMELLA LESCURII

PLEUROZIUM SCHREBERI D M Large erect pinnately branched moss with a red stem, blunt concave leaves, and no costa

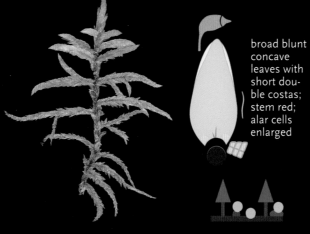

broad blunt concave leaves with short double costas; stem red; alar cells enlarged

A MAJOR BOREAL-FOREST DOMINANT that makes continuous carpets on the floors of dry forests. Immediately recognized by the erect stance, pinnate branching, red stems, and blunt concave leaves. In boreal forests, it often grows with *Ptilium* and *Hypnum,* which are more regularly pinnate and have slender curved leaves, and *Hylocomium,* which has arching, fernlike fronds and stepladder growth.

PLEUROZIUM SCHREBERI

PSEUDOCALLIERGON F R Large, brown-green, sparsely branched, wormlike mosses of limy wetlands; leaves concave, broadly oval to nearly round, with a single costa

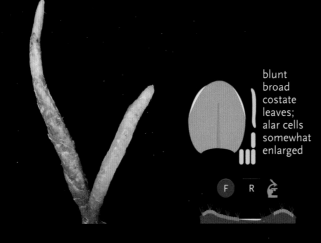

blunt broad costate leaves; alar cells somewhat enlarged

F R 🔬

MEDIUM-SIZED WORMLIKE MOSSES of rich fens and alvars, with broad, concave, closely overlapping leaves. Both species are high-northern and at their southern limits in the NFR. The species are distinct in the field but not easy. The habitat and the microscope help. *Trifarium* is a rare species of rich fens, associated with species like *Campylium stellatum, Helodium blandowii, Calliergonella cuspidata,* and *Scorpidium scorpioides.* Its marks are the cylindrical shoots and the broad blunt concave leaves with a short costa and enlarged alar cells.

PSEUDOCALLIERGON TRIFARIUM

QUICK GUIDES TO HABITATS

QUICK GUIDES TO ACROCARPS

QUICK GUIDES TO PLEUROCARPS

QUICK GUIDES TO SPHAGNUM

ACROCARPS

PLEUROCARPS

SPHAGNUM

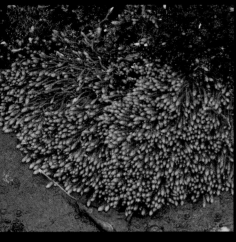

stems cylindrical and wormlike; leaves pointed; costa short and double; alar cells small

A RARE SPECIES of limy pools in flat-rock barrens, forming mats where seepage enters the pools. Also reported from fens. We have found it in association with species of *Bryum*, *Ditrichum*, and *Tortella*. Like *Pseudocalliergon trifarium*, it has cylindrical shoots and broad concave leaves. It differs in the pointed tips of the leaves, short double costa, and small alar cells. *Scorpidium scorpioides*, a fen species, looks similar but typically has longer curved leaf tips and a small group of inflated alar cells.

PSEUDOCALLIERGON TURGESCENS

PSEUDOTAXIPHYLLUM C M

Shiny somewhat flattened mosses of moist rocky places; leaves not decurrent; alar cells not differentiated; leaves toothed near the tip; brood branches often present (bb)

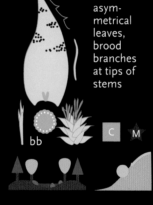

asymmetrical leaves, brood branches at tips of stems

bb

C M

TWO REASONABLY COMMON SPECIES of moist shady places: wet ledges, stream banks, dirty rocks. Both resemble *Plagiothecium*, *Isopterygiopsis*, *Taxiphyllum*, *Pylaisiadelpha* and others. Plants with brood bodies can be recognized in the field; plants without them need the microscope. The critical features are the nondecurrent leaves, undifferentiated alar cells, and stems with the outer cells thick walled.

PSEUDOTAXIPHYLLUM DISTICHACEUM

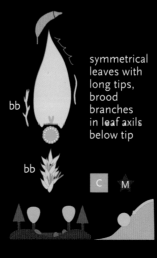

bb

symmetrical leaves with long tips, brood branches in leaf axils below tip

bb

C M

OUR TWO *PSEUDOTAXIPHYLLUM* SPECIES are similar and grow in similar places. The brood bodies are the best distinction. In *distichaceum* they are at the tips of the stems, and have leaf primordia only at their tips. In *elegans* they are in the leaf axils and have primordia along their sides. Lacking brood bodies you can use the symmetry of the leaves and length of the leaf tips. The brood bodies are better, and worth looking for.

PSEUDOTAXIPHYLLUM ELEGANS

PTERIGYNANDRUM FILIFORME Small dark oval-leaved moss making traceries and thin mats on rocks; leaves blunt and concave; cells protrude at tips

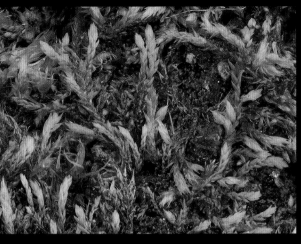

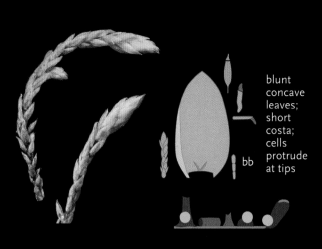

blunt concave leaves; short costa; cells protrude at tips

bb

A SMALL DARK MOSS, common but inconspicuous, on rocks in shade. Stems stringy, making thin mats; leaves concave and spoon shaped; cells long and protruding at tips; alar cells not differentiated. About the size of *Homomallium*, which also has oval leaves, but darker and much less shiny.

PTERIGYNANDRUM FILIFORME

PTILIUM CRISTA-CASTRENSIS Light-green, erect or arching, fernlike fronds; leaves slender, with long curved tips

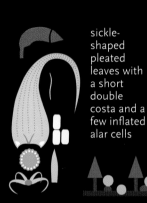

sickle-shaped pleated leaves with a short double costa and a few inflated alar cells

A COMMON MOSS of the boreal forest floor, often mixed with other species, particularly *Hypnum imponens*, *Pleurozium schreberi*, *Polytrichum commune*, *Hylocomium splendens*, and *Dicranum polysetum*. The pale-green color, long leaf tips, and regularly pinnate fronds are distinctive. *Hypnum imponens*, also pinnate, is often golden green and has shorter leaf tips and makes denser mats.

PTILIUM CRISTA-CASTRENSIS, with *Polytrichum commune*

PYLAISIA Small curly branched mosses making mats or fringes on tree bark; leaves without costas, capsules erect, alar cells small and square

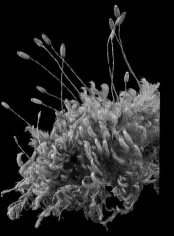

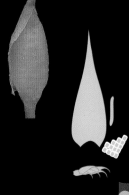

slender long-pointed ecostate leaves whose tips curl the same way; over 10 alar cells; capsules erect, with very short beaks

SMALL COMMON SHINY MOSSES of tree bark, recognized as a group by the curved branches and short-beaked erect capsules. Often with *Platygyrium*, which has straight branches, longer beaked capsules, and brood bodies. *Selwynii* seems to be our commonest species. *Intricata* and *polyantha* differ slightly in the number of alar cells, the toothing of the leaves, and the development of the peristome teeth.

PYLAISIA SELWYNII GROUP*

*According to the books, *P. selwynii* has falcate-secund leaves and over 10 square alar cells on the margins; *P. polyantha* has straight leaves and over 10 square alar cells on the margins; *P. intricata* has falcate-secund leaves and less than 10 square alar cells on the margins. There are also small differences in the development of the inner peristome. They intergrade, and it is not clear whether one is commoner than the others.

143

QUICK GUIDES TO HABITATS

QUICK GUIDES TO ACROCARPS

QUICK GUIDES TO PLEUROCARPS

QUICK GUIDES TO SPHAGNUM

ACROCARPS

PLEUROCARPS

SPHAGNUM

PYLAISIADELPHA TENUIROSTRIS

Small shiny moss without many field characters; leaves with tiny teeth, two or three alar cells inflated; slender pseudoparaphyllia (pp)

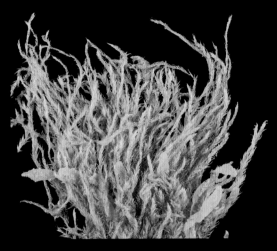

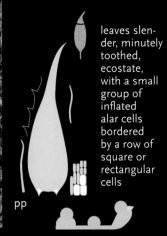

leaves slender, minutely toothed, ecostate, with a small group of inflated alar cells bordered by a row of square or rectangular cells

pp

A SMALL MOSS with shiny ecostate leaves and very little else. We see it on limy rocks; elsewhere it grows on trees. Similar to *Hypnum pallescens* in size and slender-tipped leaves. Best recognized microscopically: it has a small group of strongly inflated alar cells bordered by a larger group of rectangular cells that extend up the margin.

PYLAISIADELPHA TENUIROSTRIS

RAUIELLA SCITA

Small wiry once-pinnate moss of tree bark; paraphyllia lobed, cells with multiple papillae

long-tipped stem leaves; cells with multiple papillae on lower side; paraphyllia (p) lobed

b s p

A WIRY ONCE-PINNATE MOSS that grows on tree bark. Resembling *Abietinella* but smaller. Recognized by the size and habitat; confirmed by the single-costate leaves with long tips, lobed paraphyllia, and cells only papillose below. *Cyrto-hypnum* has simple paraphyllia and cells papillose on both sides; *Haplocladium* has single papillae.

RAUIELLA SCITA

RHYNCHOSTEGIUM SERRULATUM

Small moss of moist forest floors, with flattened shoots and toothed leaves with twisted tips

flattened shoots; slender, toothed leaves with twisted tips; alar cells weakly differentiated

A SHINY FLATTENED MOSS of deciduous forest floors, growing on soil, rocks and woods. Close to *Brachythecium*, which it resembles in the slender leaves and weak single costa that ends in a spine. Separated from *Brachythecium* by the combination of toothed edges, twisted tips, and barely differentiated alar cells. Resembles *Plagiothecium* and its relatives in the strongly flattened shoots; no *Plagiothecium* has a costa or is this toothy.

RHYNCHOSTEGIUM SERRULATUM

144

RHYTIDIADELPHUS [C] [M]

Large erect irregularly branched mosses, bristly when wet and shaggy like pipe cleaners when dry; triangular leaves with long tips and double costas

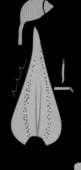
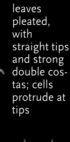

leaves pleated, with straight tips and strong double costas; cells protrude at tips

TWO LARGE MOSSES, recognized by the irregular branching and triangular leaves with double costas.* *Triquetrus* is common in swamps, stream bottoms, and moist meadows, and locally abundant in boreal forest. It can be identified on sight by its erect growth and shaggy look. The pleated leaves with strong costas are a good confirmation.

RHYTIDIADELPHUS TRIQUETRUS

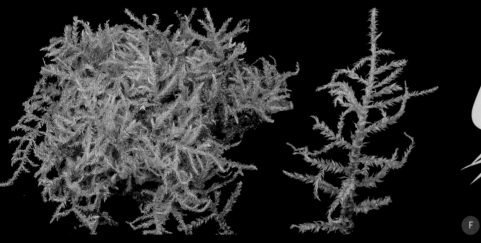

leaves with slender arching tips; costa weak; cell tips do not protrude

A SMALLER SPECIES, growing in loose tangles in swamps and on rocks in streams. The slender arching tips of the stem leaves are distinctive. Compared to *triquetrus*, the leaves have longer tips, the costas are weaker, and the cell tips do not protrude.

RHYTIDIADELPHUS SQUARROSUS

RHYTIDIUM RUGOSUM [F] [U]

Large golden moss of exposed limy soil; branches thick; leaves ripply, with long curved tips and strong single costas

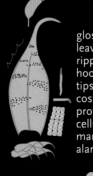

glossy leaves with ripples and hooked tips, single costas, protruding cells, and many small alar cells

A HANDSOME MOSS of open limy ledges and soils, forming dense mats and cushions on ledge crests and summits. Recognized by the bright color and ripply, hooked leaves. Confirmed by the protruding cells and numerous small alar cells. *Hypnum imponens* has similarly curved leaves but is about half the size and lacks a costa.

RHYTIDIUM RUGOSUM

*A third species, *Rhytidiadelphus loreus,* occurs rarely in the Maritimes. It has falcate-secund leaves which are pleated at their bases and undifferentiated alar cells.

QUICK GUIDES TO HABITATS

QUICK GUIDES TO ACROCARPS

QUICK GUIDES TO PLEUROCARPS

QUICK GUIDES TO SPHAGNUM

ACROCARPS

PLEUROCARPS

SPHAGNUM

SANIONIA UNCINATA Slender, loosely pinnate moss with hooked costate leaves that are pleated when dry

leaves slender, costate, pleated, often hooked; alar cells inflated; stems with a central strand and inflated cortex

A SICKLE-LEAVED WETLAND MOSS, related to *Drepanocladus* and *Warnstorfia* and like them with costate leaves and inflated alar cells. Separated from them by two main features: the pleated leaves and the inflated stem cortex. Ecologically tolerant and found in a wide range of habitats: swamps, pools, moist ledges, dryer rocks, tree trunks, and logs. Wide ranging and regularly encountered but not common in the areas we work.

SANIONIA UNCINATA

SCHWETSCHKEOPSIS FABRONIA ᴿ Small shaggy unspellable moss with loose descending branches; on tree bark and shaded limy ledges.

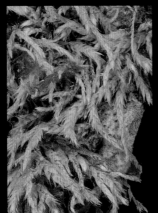

leaves long pointed, without costas; small diamond-shaped cells with protruding tips; square cells up leaf edges

A SMALL MOSS of tree bark with a southern distribution, barely reaching the NFR. Nondescript in the field: light green, silky-looking, with long-tipped leaves and loose descending branches. Well-marked microscopically: the combination of ecostate, long-tipped leaves and diamond-shaped cells that protrude at their tips is unique.

SCHWETSCHKEOPSIS FABRONIA

SCORPIDIUM SCORPIOIDES ꜰ ᴿ Large moss of limy fens; cylindrical shoots; concave leaves with short costas and inflated alar cells; leaf tips secund; branch tips often hooked

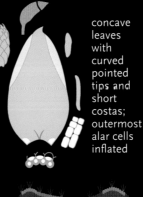

concave leaves with curved pointed tips and short costas; outermost alar cells inflated

THE BLACK BOG MOSS OF THE NORTH, covering large expanses of northern fens. Large plants, often dark brown or red-black, with cylindrical shoots and sparse branching. Leaves concave, with short curved tips; a few of the outermost alar cells inflated. *Calliergonella cuspidata* can look similar, especially when young. It is usually more branched, has blunter leaf tips that are never secund, and has inflated alar cells that extend almost to the costa.

SCORPIDIUM SCORPIOIDES

SEMATOPHYLLUM MARYLANDICUM [U] 🔬

Medium-sized golden-green moss on rocks by streams; costa short; alar cells inflated; capsules beaked

capsules with a beak; leaf margins not incurved or toothed; alar cells abruptly inflated; stems without a central strand; cells of capsule with rounded corners

AN UNCOMMON MOSS of rocky streams and waterfalls, restricted to the Appalachians and absent from the Great Lakes. Recognized, approximately, by the golden color, concave leaves, and fat branches. Confirmed by the long-beaked capsules, abruptly inflated alar cells, and capsule cells with rounded corners. *Hygrohypnum eugyrium* is very similar and has to be excluded carefully. It has incurved leaf margins, wider shoulders to the leaves, secund leaf tips, and capsules with shorter beaks. Under the microscope, its stems have a central strand and its capsule cells have sharp corners.

SEMATOPHYLLUM MARYLANDICUM

TAXIPHYLLUM DEPLANATUM [U] 🔬

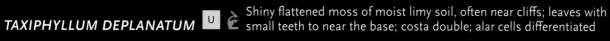

Shiny flattened moss of moist limy soil, often near cliffs; leaves with small teeth to near the base; costa double; alar cells differentiated

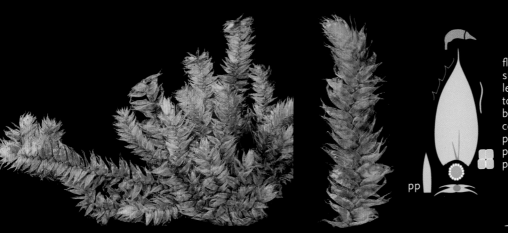

flattened shoots; leaves toothed to base; alar cells square; pseudo-paraphyllia present (pp)

pp

A SHINY MOSS with flattened shoots and overlapping leaves, similar to *Plagiothecium* and once part of it. It is regularly associated with limy seepage, and may be locally common where the habitat is right. The best characters are microscopic: *Taxiphyllum* has lanceolate pseudoparaphyllia (pp) and toothed nondecurrent leaves with small groups of square alar cells.

TAXIPHYLLUM DEPLANATUM

THAMNOBRYUM ALLEGHANIENSE [C] [M]

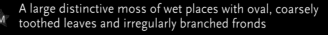

A large distinctive moss of wet places with oval, coarsely toothed leaves and irregularly branched fronds

oval leaves with sharp teeth, strong costas, and small, diamond-shaped cells

A BIG SHAGGY FRONDOSE MOSS, with an Appalachian distribution, common on rocks in the spray zone of streams, especially in gorges and near waterfalls. Also on stream banks and in cracks in cliffs. The largest NFR populations may be in the shut-in gorges of the Catskills and Helderbergs. Related to *Neckera* and branches in much the same way. The oval leaves with jagged teeth and flattened, loosely pinnate sprays of branches are distinctive.

THAMNOBRYUM ALLEGHANIENSE

QUICK GUIDES TO HABITATS

QUICK GUIDES TO ACROCARPS

QUICK GUIDES TO PLEUROCARPS

QUICK GUIDES TO SPHAGNUM

ACROCARPS

PLEUROCARPS

SPHAGNUM

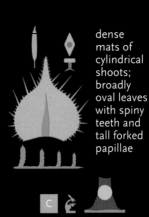

dense mats of cylindrical shoots; broadly oval leaves with spiny teeth and tall forked papillae

C

A DISTINCTIVE GENUS of small papillose mosses that make dense mats on the bases and bark of trees in the oak zone. The leaves are broadly oval and densely packed, making cylindrical shoots; their teeth are long and spiny. We have two species: *asprella* with branched papillae and *hirtella* with simple ones. *Asprella*, shown here, is reasonably common in the southern NFR. *Hirtella* seems a bit rarer. Both drop out northwards and in the mountains. Compare *Myurella julacea*, also cylindrical and papillose but without spiny teeth and restricted to sites with limy seepage.

THELIA ASPRELLA C M

THUIDIUM C M Large mosses making fernlike fronds with twice-pinnate branching; cells with single papillae over the centers; branched paraphyllia (p) on stems, making them look fuzzy*

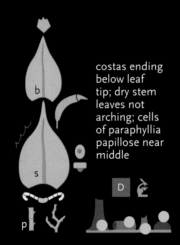

costas ending below leaf tip; dry stem leaves not arching; cells of paraphyllia papillose near middle

D

LARGE FRONDOSE MOSSES common on the forest floor, also on boulders, tree bases, logs, and ledges. The regularly twice-pinnate, fernlike fronds are unique; *Hylocomium splendens*, the only thing that comes close, is three-times pinnate. The leaves are small, oval, and opaque. The cells have single papillae, and the stems have a dense woolly covering of paraphyllia.

The species are close. *Delicatulum* is common on rocks, soil, tree bases, and logs in moist woods, often forming large patches. It has short costas that don't reach the tips of the stem leaves, and paraphyllia with the papillae near the middle of the cells.

THUIDIUM DELICATULUM

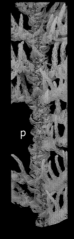
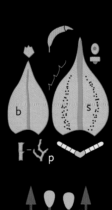

costas filling the tips of the stem leaves; cells of paraphyllia papillose near ends

C

A CLOSELY RELATED SPECIES, found in limy swamps and on limy or sub-limy ledges. Sometimes recognizable in the field by the strong costas that reach the tips of the stem leaves. Confirmed in the lab by the strong costas and paraphyllia whose cells are papillose at their ends.

THUIDIUM RECOGNITUM

*The once-pinnate plants that were formerly in *Thuidium* are now in *Abietinella*, *Cyrto-hypnum*, and *Rauiella*. *Heterocladium* and *Haplocladium* are two other once-pinnate genera that resemble *Thuidium*.

TOMENTYPNUM NITENS C M Large species of rich fens with pinnate branching and long straight slender-tipped leaves

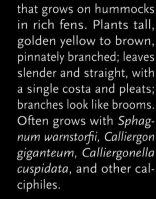

slender sharp-tipped, pleated leaves; porose basal cells; alar cells not inflated

A LARGE DISTINCTIVE MOSS that grows on hummocks in rich fens. Plants tall, golden yellow to brown, pinnately branched; leaves slender and straight, with a single costa and pleats; branches look like brooms. Often grows with *Sphagnum warnstorfii*, *Calliergon giganteum*, *Calliergonella cuspidata*, and other calciphiles.

*TOMENTYPNUM NITENS**

TORRENTARIA RIPARIOIDES* C M Large stringy moss found on wet rocks in brooks; oval leaves with a single costa, small teeth, and no border

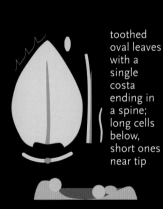

toothed oval leaves with a single costa ending in a spine; long cells below, short ones near tip

A COMMON MOSS in brooks, growing low on rocks, submerged at high water. Recognized by the loosely arranged, oval, costate leaves that spread widely when wet. *Platylomella* can be similar but is smaller and has bordered leaves. Some Hygrohypnums have similar leaves; their leaves are typically less spreading and either have double costas or differentiated alar cells.

*TORRENTARIA RIPARIOIDES**

WARNSTORFIA C Aquatic mosses with long, slender, often hooked leaves; costa single, with rhizoid initials near the tips; alar cells enlarged

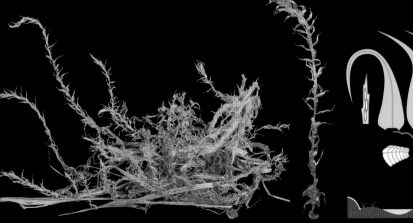
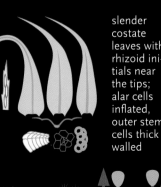

slender costate leaves with rhizoid initials near the tips; alar cells inflated, outer stem cells thick walled

SICKLE-LEAVED COSTATE MOSSES formerly included in *Drepanocladus*, and like it with inflated alar cells that extend to the costa and small thick-walled cells in the stem cortex. Separated from *Drepanocladus* by the large clear cells—rhizoid initials—near the stem tips. Our two species, *fluitans* and *exannulata*, are highly variable and usually separated by differences in the development of the alar cells. We don't have much faith in the distinction and consider them a group.

WARNSTORFIA FLUITANS GROUP

* *Torrentaria*, a distinctive moss of uncertain affinities, is also called *Rhynchostegium aquaticum*, *Platyhypnidium riparioides*, and *Eurynchium riparioides*. *Torrentaria* is the prettiest. *Tomentypnum* is also spelled *Tomenthypnum*. As in *Helodium*, some authors use a *h* to indicate the Greek rough breathing and some don't. The anti-*h* faction seems to be prevailing.

149

QUICK GUIDES TO HABITATS

QUICK GUIDES TO ACROCARPS

QUICK GUIDES TO PLEUROCARPS

QUICK GUIDES TO SPHAGNUM

ACROCARPS

PLEUROCARPS

SPHAGNUM

SECTION *ACUTIFOLIA* C M Green cells widest on concave (upper) sides of branch leaves (◯◯); plants often with red or deep brown color; many species recognizable in the field*

round-topped and round-headed; stem often red; stem leaves inrolled at the tip, branch leaf pores large**

D M

THE ACUTIFOLIA are a large group, common in wet woods and dominant in many peatlands. Some species are identifiable some of the time by their color or stem leaves; all are identifiable by the trapezoidal green cells in the branch leaves that are widest on the upper side of the leaf. *Capillifolium* is a very common species, found as pink and red plants in bog hummocks and as green plants on ledges and in swampy woodlands. The small round heads and large stem leaves with inrolled tips are good marks.

SPHAGNUM CAPILLIFOLIUM

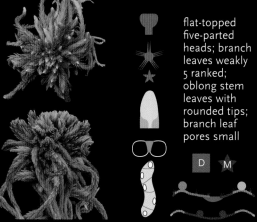

flat-topped five-parted heads; branch leaves weakly 5 ranked; oblong stem leaves with rounded tips; branch leaf pores small

D M

RUBELLUM IS A PEATLAND DOMINANT that grows in lawns and on the sides of hummocks. It differs from *capillifolium* in the flat-topped heads, more rounded stem leaves, and weakly five-ranked branch leaves with small pores near the tip. Sun plants are deep red, shade plants green or mottled. Can grow with *warnstorfii* in fens and have pores that are almost as small. Intermediates between *rubellum* and *capillifolium* are easy to find. You can call them *subtile*, but it doesn't really help.

SPHAGNUM RUBELLUM

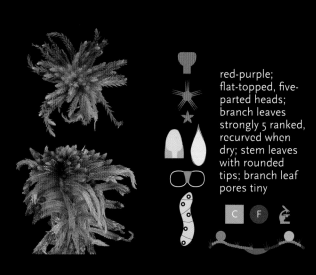

red-purple; flat-topped, five-parted heads; branch leaves strongly 5 ranked, recurved when dry; stem leaves with rounded tips; branch leaf pores tiny

C F 🔬

A PURPLE-RED SPECIES of hummocks in limy fens, with strongly five ranked branch leaves with tiny ringed pores, only a couple of microns wide, near the tips. Dry plants look glossy or metallic and have spreading or recurved branch leaves. *Rubellum*, which can grow in fens, is quite similar: it is less purple, less five ranked, and has larger pores.

SPHAGNUM WARNSTORFII

* The rare coastal species *bartlettianum* and *molle* are not shown. The common woodland plants, sometimes called *S. subtile*, that are intermediate between *capillifolium* and *rubellum* are *capillifolium* to us. The mosses on these two pages—the "red *Acutifolia*"—are common and variable. Often you call tell them apart with just a lens. Equally often you need a microscope. Sometimes, as elsewhere in *Sphagnum*, you can't tell them apart at all.

SECTION *ACUTIFOLIA*

plants red or mottled red and green; heads five parted; branch leaves not strongly ranked; stem leaves with a rounded tip with a ragged center; some outer stem cells with pores

A WEAKLY CHARACTERIZED SPECIES, often found on pond shores or the edges of wet mats, and often mottled red and green. Close to *rubellum*, from which it is said to differ in the unranked branch leaves, stem leaves with teeth or lacerations at their tips, and outer stem cells with a few pores. Possibly less a species than a group of transitional forms. We often see plants that look like *russowii* in the field but lack the microscopic characters it is supposed to have. Sphagnums are like that. Dick Andrus says in the *Flora of North America* that it is "not particularly distinct phenotypically" and our "most frequently misidentified *Sphagnum* species." We say there is such a thing as trying too hard.

SPHAGNUM RUSSOWII

SECTION *ACUTIFOLIA*: BROWN-STEMMED SPECIES

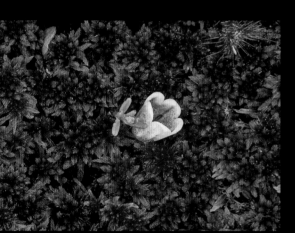

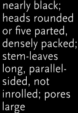

heads deep brown; stems nearly black; heads rounded or five parted, densely packed; stem-leaves long, parallel-sided, not inrolled; pores large

A VERY COMMON SPECIES of bog hummocks, usually growing higher than any other *Sphagnum*; can cover large areas in dry bogs. The dark color and small, densely packed heads are good characters. In coastal bogs, often mixed with *S. flavicomans*, which typically has larger heads and is more orange brown. The two are close and don't separate cleanly.

SPHAGNUM FUSCUM (brown plants, with *Sphagnum rubellum*)

orange brown, with dark stems and rounded or five parted heads; stem leaves taper to an inrolled tip; pores large

A LARGE, HUMMOCK-FORMING SPECIES, common on the Atlantic coast, mostly absent inland. Plants dark, resembling *S. fuscum* and often mixed with it, separated in the field by their larger, rounder, looser and more orange-brown heads. Separated in the lab, in theory, by the shorter branch leaves and more tapered and inrolled stem leaves. In practice, as Lewis Anderson says in *Maine Mosses*, it is often "more recognizable in the field than under the microscope."

SPHAGNUM FLAVICOMANS, with *Sphagnum rubellum*; often darker than the picture

** The icons in this section typically show the head in profile 🍄 and from above ★ ; the stem and branch leaves ▬ ◗ ; the leaf cross-section with the concave side up ∞, showing a green cell flanked by two clear cells; and the outside and inside surfaces of the clear cells of the branch leaves

QUICK GUIDES TO HABITATS

QUICK GUIDES TO ACROCARPS

QUICK GUIDES TO PLEUROCARPS

QUICK GUIDES TO SPHAGNUM

ACROCARPS

PLEUROCARPS

SPHAGNUM

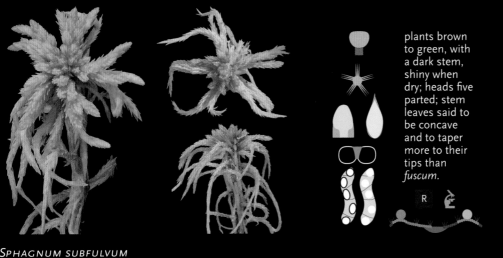

plants brown to green, with a dark stem, shiny when dry; heads five parted; stem leaves said to be concave and to taper more to their tips than *fuscum.*

A BROWN-GREEN SPECIES of hummocks in fens with flat-topped, five-parted heads, a dark stem, and stem leaves rounded at their tips. Weakly separated from *fuscum* and *flavicomans*: Lewis Anderson said in *Maine Mosses* that it is "difficult to define." Depending on the author, its critical features are said to be its light color, shininess when dry, strongly inrolled branch leaf tips, strongly bulging clear cells, and more concave and tapering stem leaves. None of these characters are reliable, and specimens that seem to be this in the field may run to other species under the microscope.

SPHAGNUM SUBFULVUM

SECTION *ACUTIFOLIA:* GREEN SPECIES

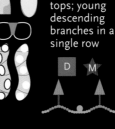

flat heads, branches stiffly arranged in five rays; stem leaves oblong, fringed across the tops; young descending branches in a single row

A COMMON SHADE SPECIES, covering large areas in forested conifer swamps and wet conifer forests. Also common on shaded ledges. Heads five parted, the branches radiating stiffly; stem leaves erect, fringed across the top, making a cup at the tip of the stem after the head is removed. *Sphagnum fimbriatum* is similar but has broader stem leaves and is often in the open. *Sphagnum fallax* resembles *girgensohnii* and can mix with it. It has small triangular stem leaves without fringes and young descending branches in two rows.

SPHAGNUM GIRGENSOHNII

heads irregularly five parted, with a large central bud; stem leaves fan shaped, flaring upwards, fringed across the tip and down the sides

A COMMON SPECIES making cushions or small carpets on pond shores and at the wet edges of floating mats; also on wet ledges and in hollows in forests. Light white green, not strongly five parted, often with capsules. Stem leaves fan shaped and fringed, making a cup or a crown. *Girgensohnii*, also fringed, is more strongly five parted and has narrower stem leaves.

SECTION *ACUTIFOLIA*: GREEN SPECIES

heads dense and round; branches in clusters of five or more, with three spreading branches in each cluster; stem leaves small, pointed

U F

A NORTHERN SPECIES of cold forests, commonest in the mountains and near the coast but nowhere abundant. Plants small, with dense round heads, resembling the shade forms of *capillifolium*. The branches have weakly five-ranked leaves, and at least some of the branch clusters have three spreading branches. Typically found on shaded cliffs, often with seepage. Also on dirty boulders and the forest floor.

SPHAGNUM QUINQUEFARIUM

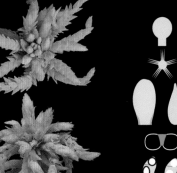

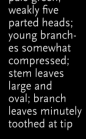

pale-green, weakly five parted heads; young branches somewhat compressed; stem leaves large and oval; branch leaves minutely toothed at tip

F U

A COASTAL SPECIES of medium fens and wet bogs with some water flowing through. Grows low to the water, often in wet mats where it is hard to walk. Plants pale green or yellow; stem leaves large. Tips of branch leaves minutely toothed under the microscope. A fairly distinctive species but not strongly marked in the field. The pale color, short thick branches, and somewhat flattened young branches are the best characters.

SPHAGNUM ANGERMANICUM, yellow-green, with *S. magellanicum*, red

SECTION *CUSPIDATA* 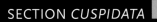 Green and brown Sphagnums with small pores, short stem leaves, and trapezoidal green cells widest on the outside of the leaf; many species recognizable in the field

limp aquatic, collapsing out of water; pale green or yellow; leaves long and inrolled to needle tips; stem leaves pointed*

C M

COMMON MOSSES of bogs, fens, and forests, always in loose wet mats. Green, light brown, or orange but never red or deep brown. Recognized microscopically by the characters given above. Identifiable species-by-species in the field but not as a group. *Cuspidatum* is a common species of bogs and fens, found floating in pools or sprawled on the mud when the pools dry. Its marks are its light color, long inrolled leaves, and limpness. The short, pointed stem leaves and cells with few pores are good lab characters.

SPHAGNUM CUSPIDATUM

* The ⬛ icon represents the cross section of a branch leaf with a trapezoidal green cell, widest on the convex (outside) side of the leaf, flanked by two clear cells.

QUICK GUIDES TO HABITATS

QUICK GUIDES TO ACROCARPS

QUICK GUIDES TO PLEUROCARPS

QUICK GUIDES TO SPHAGNUM

ACROCARPS

PLEUROCARPS

SPHAGNUM

SPHAGNUM MAJUS

stiff dark-colored plants, growing erect; leaves long and inrolled to needle tips; pores of leaf cells in center

C M

A COMMON PLANT of low wet bog mats, filling depressions and making extensive lawns. Often dirty green or blackish; leaves with long inrolled tips like those of *cuspidatum* but the stem leaves rounded and the plants darker and more erect. *Torreyanum* also has long inrolled leaves; it is larger than *majus*, often very dark, and has a bristly look especially when dry.

SPHAGNUM TORREYANUM

plants large and stiff; heads weakly five parted; leaves with long tips, recurved when dry; stem leaves rounded; green cells triangular

F U

A LARGE SPECIES with long slender leaf tips, found in low wet bog and fen mats and in bog pools, mostly near the coast. In the field recognized by the large size, dark-green or black color, and the long-tipped leaves which give it a bristly look. Confirmed by the triangular green cells and leaf cells with relatively few pores. *Majus* is smaller, less bristly, and has a row of pores down the middle of the cells.

yellow brown or orange brown; five-parted heads and branch leaves strongly five ranked; green cells don't reach inner surface of leaf

D M

A DISTINCTIVE SPECIES of coastal bogs and fens growing in low wet mats and covering large areas. Recognized by the bright color, five-parted heads, and five-ranked branch leaves that give the branches sawtooth edges. Confirmed by the green cells which are exposed only on the outer surfaces of the leaves.

SECTION *CUSPIDATA*

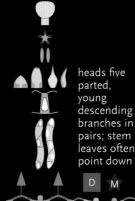

heads five parted, young descending branches in pairs; stem leaves often point down

D M

Very common Sphagnums, on hummocks in bogs and in large wet carpets in bogs, swamps, and poor fens. Green in the shade and yellow or light brown in the sun; heads always five parted, almost always with the young descending branches paired. Dry leaf tips often wavy or hooked. The group is divided into four species by the shape of the stem leaves, the size of the plants, and the arrangement of the branch leaves. We have tried to use these characters for years; they do not work for us.

*Sphagnum recurvum group**

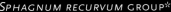

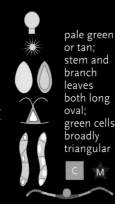

branch   stem

pale green or tan; stem and branch leaves both long oval; green cells broadly triangular

C M

A small *Sphagnum*, common in bog pools and wet bogs and fens in the north and near the coast. Light green or brown, soft or limp, often submerged. Recognized by the oval stem leaves, similar in size and shape to the branch leaves. Confirmed by the small, broadly triangular green cells and tall bulging clear cells.

Sphagnum tenellum

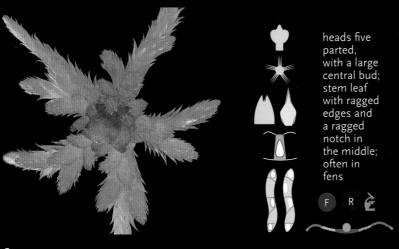

heads five parted, with a large central bud; stem leaf with ragged edges and a ragged notch in the middle; often in fens

F R ℰ

A common species in the subarctic, rare with us. Associated with wet, somewhat minerotrophic fens and pools in bogs. Resembles the *recurvum* group and, like them, grows in hollows and lawns near the water table. Recognized in the field by the pale color, five parted heads, large terminal bud, and, with a good lens, the split down the middle of the stem leaves.

Four other northern species of Section *Cuspidata*, not shown here, occur rarely in the NFR. *Lindbergii* is a wide-ranging peatland species with a dark stem and broad stem leaves fringed across their summits, resembling those of *Sphagnum fimbriatum*. *Jensenii* and its sister species *annulatum* are large aquatic species, resembling *majus* but with small heads and more numerous pores. *Balticum* is a member of the *recurvum* group with squarrose stem leaves and branch clusters with a single descending branch.

Sphagnum riparium

* Including *S. recurvum, angustifolium, fallax,* and *flexuosum.* The icon ‖ means that the young descending branches are paired. See the pictures

QUICK GUIDES TO HABITATS

QUICK GUIDES TO ACROCARPS

QUICK GUIDES TO PLEUROCARPS

QUICK GUIDES TO SPHAGNUM

ACROCARPS

PLEUROCARPS

SPHAGNUM

SECTION POLYCLADA Large species of fertile conifer swamps; plants big-headed and free-standing, on hummocks in swamps, looking like palm trees or lollypops; 6 or more branches per cluster

large, mop-headed, branches
in clusters of six or more

A SHAGGY BIG-HEADED SPECIES of fertile conifer swamps. Found throughout the NFR, most commonly under white cedar. The driest and least colonial of our Sphagnums, often growing as scattered plants, sometimes in loose mounds. The lollypop heads and brown stems with six or more branches per cluster are diagnostic.

SPHAGNUM WULFIANUM, with *Hylocomium splendens*

SECTION RIGIDA F U Short plants forming dense mounds; leaf tips inrolled and slightly hooded; stem leaves much shorter than branch leaves

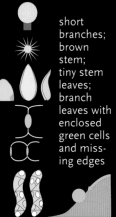

short branches; brown stem; tiny stem leaves; branch leaves with enclosed green cells and missing edges

A WIDE-RANGING SPECIES of acid wet sand and wet cliffs, recognized by the short branches and dense heads. Leaves with elongate, blunt, inrolled tips, like gravy boats. Looks something like a small version of one of the species in Section *Sphagnum* and, like them, has the leaf edges missing. Best confirmed by the short, rounded stem leaves.*

SPHAGNUM COMPACTUM

SECTION SPHAGNUM D M Large mosses with short fat branches and hooded branch leaves whose edges are partly missing. The section is identifiable in the field; except for *magellanicum,* the species are not.

red or green, with thick branches and hooded leaves; green cells completely enclosed by clear cells

A COMMON GROUP of large mosses recognized by the thick branches with closely overlapping leaves and the leaves with deeply hooded tips. For confirmation, the branch-leaf edges are missing (OC) and the stem cortex has fibrils. *Magellanicum* is a very common species, found in all sorts of wetlands: bogs, poor fens, rich fens, and wooded swamps. Open-grown plants are red and identifiable in the field. Shade plants are green and identified by sectioning the branch leaves to see the enclosed green cells (◌).

SPHAGNUM MAGELLANICUM

* *Sphagnum strictum* is a related species, southern and rare in the NFR, that is distinguished from *compactum* microscopically.

SECTION *SPHAGNUM*

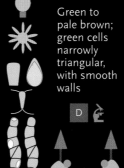

Green to pale brown; green cells narrowly triangular, with smooth walls

SPHAGNUM PALUSTRE

THE COMMON *SPHAGNUM* of wet woods and swamps, found in hollows, around pools, and in seeps and wet depressions. Also in fens and marshes. A large species with fat branches and sometimes, as in the left picture, squarrose leaves. Identified by leaf sections: the green cells are narrowly triangular and have smooth walls ( ).

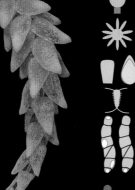

pale brown to rich golden brown; short fat branches; clear cells with papillose walls

SPHAGNUM PAPILLOSUM

A DOMINANT SPECIES of poor to medium fens and wet bog mats, often covering large areas. Less common in raised bogs, where it is replaced by *magellanicum*. Often turns a rich golden brown in late summer. Identified by leaf sections: the green cells are narrowly triangular and have papillose walls.

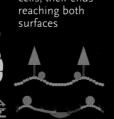

green or golden brown, with short fat branches; green cells centered between clear cells, their ends reaching both surfaces

SPHAGNUM CENTRALE

A COMMON SPECIES in fertile conifer swamps and rich fens, light green in the shade, golden brown in the open. Identified by leaf sections: the green cells are slender and oval, with ends that reach both surfaces ( ). Be careful with *magellanicum*, which has green cells that are completely enclosed, without the extended ends.

QUICK GUIDES TO HABITATS

QUICK GUIDES TO ACROCARPS

QUICK GUIDES TO PLEUROCARPS

QUICK GUIDES TO SPHAGNUM

ACROCARPS

PLEUROCARPS

SPHAGNUM

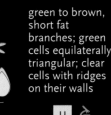

green to brown, short fat branches; green cells equilaterally triangular; clear cells with ridges on their walls

U

An UNCOMMON SPECIES of mineral-rich waters, especially near the coast: fens, marshes, shrub swamps, forested swamps, pond shores. Identified by leaf sections: the green cells are nearly equilateral, and the clear cells adjacent to them have lengthwise ridges—comb fibrils—on their walls (▼). We treat the species inclusively, in the old carefree sense. It may be further split, based on the presence of fibrils in the inner stem cortex, into *S. affine* and *S. austenii.*

SPHAGNUM IMBRICATUM

SECTION *SQUARROSA* C M

Minerotrophic species with oblong stem leaves fringed at the ends and branch leaves with large pores; heads with conspicuous terminal buds; both species recognizable in field

leaves abruptly narrowed to a long tip that points outward; branch leaves with large ringed pores on the inner surface

C M

A SHOWY SPECIES of fertile wooded swamps, with abruptly narrowed leaf tips that bend backwards and outwards. The large pores and oblong stem leaves with teeth across the tip are good confirmations. *S. palustre* often has hooked leaves, but they are never narrowed to slender tips like *squarrosum.*

SPHAGNUM SQUARROSUM

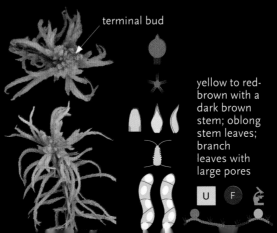

terminal bud

yellow to red-brown with a dark brown stem; oblong stem leaves; branch leaves with large pores

U F

A CHARACTERISTIC SPECIES of hummocks in medium fens and floating bogs. Recognized in the field by the yellow-green to rich red-brown color, the dark stem, conspicuous terminal bud, and large oblong stem leaves. Confirmed by the large branch-leaf pores. None of the other golden-brown or red-brown species with which it might be confused have both oblong stem leaves and a conspicuous terminal bud

SECTION *SUBSECUNDA: SUBSECUNDUM* GROUP

Yellow-brown, loose-leaved mosses with small pores, like beads, around the edges of the clear cells

 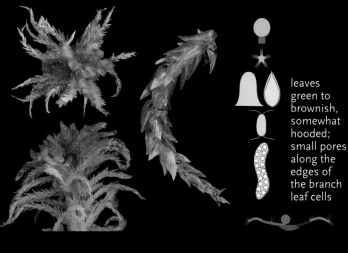

leaves green to brownish, somewhat hooded; small pores along the edges of the branch leaf cells

A COMPLICATED GROUP of closely related species. Recognized, often with difficulty, by a shaggy look, weakly five-parted heads with short, curved branches, and somewhat hooded leaves. Confirmed, easily, by the small beadlike pores. A minerotrophic group, reasonably common in fens and marshes, on pond shores, in wet spots in the woods, and on shaded, wet, mineral soil.

SPHAGNUM SUBSECUNDUM GROUP*

SECTION *SUBSECUNDA: PYLAESII*

A deep-brown oddball with stubby branches and almost no head

  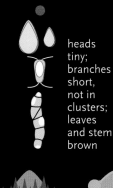  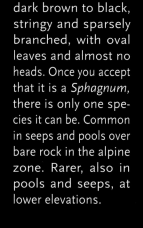

heads tiny; branches short, not in clusters; leaves and stem brown

A UNIQUE SPECIES, dark brown to black, stringy and sparsely branched, with oval leaves and almost no heads. Once you accept that it is a *Sphagnum*, there is only one species it can be. Common in seeps and pools over bare rock in the alpine zone. Rarer, also in pools and seeps, at lower elevations.

SPHAGNUM PYLAESII, with *Polytrichum juniperinum*

 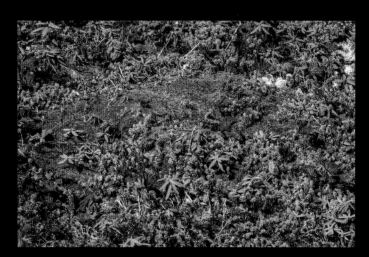

Sphagnum-sedge-dwarf shrub heath in northwestern Newfoundland; *Sphagnum fuscum* and *rubellum*, *Eriophorum*, *Trichophorum*, crowberry.

* Including *subsecundum*, *contortum*, *lescurii*, *inundatum*, and *platyphyllum*.

QUICK GUIDES TO HABITATS

QUICK GUIDES TO ACROCARPS

QUICK GUIDES TO PLEUROCARPS

QUICK GUIDES TO SPHAGNUM

ACROCARPS

PLEUROCARPS

SPHAGNUM

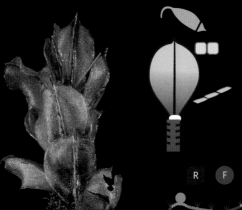

broad oval leaves with a strong border and short point; old stems and leaves blackish; border 1 cell thick; rhizoids in lines on the stem

THREE RARE SPECIES, found too late to go in the systematic sections or guides. *Cinclidium stygium* is an acrocarp with oval, strongly bordered, untoothed leaves, found in wet limy fens and seeps. It is transcontinental in the north and rare along the northern edge of the NFR. It resembles *Rhizomnium*, from which it is separated by the strong points on the leaves, the one-cell-thick leaf borders, the shiny black older leaves, and the rhizoids that originate in lines along the stem.

CINCLIDIUM STYGIUM

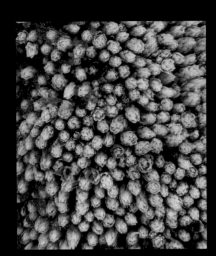

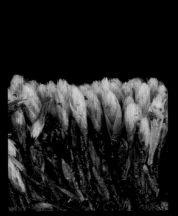

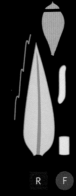

tiny, in dense clumps, bright white green or blue green; leaves under 0.8 mm long; capsules erect, with a single peristome

A TINY, BRIGHT GREEN ACROCARP with a widely scattered distribution, found in North America from the central United States to the arctic. It grows on metal-rich rocks, often in seepage, often near copper or iron deposits. Our specimen came from weathered basalts in northern Newfoundland. Its best marks are the dense clumps, the bright blue-green color, the tiny leaves, and, when present, the erect capsules, and the single peristome. *Pohlia* is vegetatively similar but considerably bigger and never blue green.

*MIELICHHOFERIA ELONGATA**

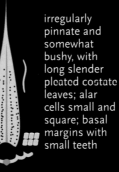

irregularly pinnate and somewhat bushy, with long slender pleated costate leaves; alar cells small and square; basal margins with small teeth

HOMALOTHECIUM is a genus of pleurocarps, close to *Bracythecium* but with narrow, pleated leaves, common in western North America and rare or absent in eastern North America. *Sericeum* is a medium-sized bushy species, widely distributed in Eurasia and Africa and known in North America from scattered sites in Newfoundland. Our collection came from a peaty coastal crowberry heath; it is also reported from forests and limestone barrens. Separated from *Tomentypnum*, which the leaves much resemble, by the less regularly pinnate shoots, more differentiated alar cells, and teeth on the lower margins.

HOMALOTHECIUM SERICEUM

* Formerly called *M. mielichhoferiana*; that name is now restricted to a species that has not been found in eastern North America

Basalt cliffs, Cape Onion, NL. *Cinclidium stygium, Distichium capillaceum, Gymnomitrion concinnatum, Mielichhoferia elongata, Pogonatum, Syntrichia ruralis.*

Anorthosite cliffs with calcareous seepage, Keene Valley, NY. *Anomobryum julaceum, Amphidium, Myurella julacea, Plagiopus oederiana, Gymnostomum aeruginosum, Campylium chrysophyllum.*

Rich fen, Montpelier, VT. *Campylium stellatum, Calliergon giganteum, Calliergonella cuspidata, Hamatocaulis vernicosus, Pseudocalliergon trifarium, Sphagnum warnstorfii, S. subsecundum.*

Carex lasiocarpa fen, near Whaleback, ME. *Sphagnum fuscum, S. girgensohnii, S. rubellum, S. recurvum, S. subsecundum, S. teres.*

Large patterned peatland, Adirondacks, NY. *Polytrichum strictum, Dicranum scoparium, Sphagnum capillifolium, S. fuscum, S. rubellum, S. recurvum*

Peaty beaver meadows, Sunderland, VT. *Aulacomnium palustre, Dicranum scoparium, Sphagnum fuscum, S. majus, S. magellanicum, S. recurvum.*

Wooded creek, White Creek, NY. *Anomodon attenuatus, Atrichum undulatum, Brachythecium plumosum, Climacium dendroides, Dicranum fulvum, Hygrohypnum duriusculum, Torrentaria riparioides.*

Stream channel in boreal forest, Fundy National Park, NB. *Bazzania trilobata, Dicranum majus, Hylocomium splendens, Hypnum imponens, Pleurozium schreberi, Polytrichastrum alpinum*

Limestone barrens with shrubby tundra, Northern Peninsula, NL. *Bryum, Distichium capillaceum, Ditrichum flexicaule, Encalypta rhaptocarpa, Tortella fragilis, T. tortuosa.*

Alvar limestone barrens, Manitoulin Island, ON. *Ditrichum flexicaule Pseudocalliergon turgescens, Tortella inclinata, T. tortuosa*

Marble knoll, Manchester, VT. *Anomodon attenuatus, A. rostratus, Encalypta procera, Myurella julacea, Plagiomnium rostratum, Schistidium crassithecium, Seligeria, Tortella inclinata, T. tortuosa*

Subalpine thickets and tundra, Mt. Mansfield, VT. *Andreaea, Bryum Cynodontium tenellum, Pogonatum, Pohlia nutans, Polytrichum juniperinum, P. piliferum, Sphagnum capillifolium*

Lakeshore boulders, Adirondacks, NY. *Andreaea, Dicranum montanum, Grimmia cf. donniana, Hedwigia ciliare, Schistidium, Racomitrium venustum.*

Log bog, Adirondacks, NY. *Bryum pseudotriquetrum, Hypnum lindbergii, Dicranum flagellare, D. montanum, Polytrichum juniperinum, Ptilidium ciliare.*

Back-dune basin, Port Joli, NS. Partly buried carpet of *Bryum, Dicranum scoparium,* and *Pleurozium schreberi,* stabilizing the open sand.

Granitic shores and boulders, Great Wass Island, ME. *Aulacomnium palustre, Calliergon stramineum, Dicranum flagellare, Schistidium maritimum, Sphagnum fimbriatum, Ulota phyllantha.*

White cedar swamp on Jewett Brook, VT. *Bazzania trilobata, Hylocomium splendens, Pleurozium schreberi, Sphagnum girgensohnii, S. wulfianum.*

Boreal forest floor, Steuben, ME. *Bazzania trilobata, Dicranum ontariense, D. polysetum, D. scoparium, Hylocomium splendens, Hypnum imponens, Polytrichum commune.*

TOOLS, SOURCES, AND PHOTOGRAPHY

ESSENTIALS First and most important: you need a hand lens. We typically carry two on a single string: a 7× Bausch and Lomb Hastings triplet and an Iwamotu 20× achromat. We use the 7× most of the time; you can start with that and add the 20× later. Once you do, you won't look back.

LIGHTS turn out to be almost as important in moss study as lenses. We like LED head lamps in the field, and LED flashlights and light panels as desk, photo, and microscope lights. The ones with 18650 batteries last for hours on a single charge. There are several good brands; Fenix head lamps and flashlights have worked well for us. Pull the head lamp down till your lens focuses the beam and you have a field microscope resting on your nose.

MICROSCOPES are not necessary when you start but will become so if you continue. Finding mosses and making reasonable field identifications does not require a microscope. Verifying these identifications, dealing with hard species, and enjoying all the other neat things that you can see with a microscope does.

For moss work a scope needs to have good achromatic optics up to 40× and a good condenser and illuminator. Older research-grade or medical-grade scopes work fine. Most student scopes don't. We buy older scopes that have been restored by reputable dealers, throw the illuminators away, and retrofit them with LED flashlights. The bottom line is that microscopes are great fun and sooner or later you will need one.

BOOKS are also fun. We learned from Howard Crum and Lewis Anderson, *Mosses of Eastern North America* (Columbia University Press, 1981) and Robert Ireland, *Moss Flora of the Maritime Provinces* (National Museum of Canada, 1982). Both are works of intelligence and love. The modern synthesis, scholarly and terse, is the *Flora of North America*, Volumes 27-28: *Bryophyta* (Oxford University Press, 2007, 2014). Our new favorites are Bruce Allen, *Maine Mosses* (New York Botanical Garden Press, two volumes, 2005, 2014); and Jean Faubert, *Flore des bryophytes du Quebec-Labrador* (Sociéty québécoisé du bryologie, 2012-2014). They are ambitious and fascinating; we use them constantly.

For recent field guides: Karl McKnight et al., *Common Mosses of the Northeast and Appalachians* (Princeton University Press, 2013), and Ralph Pope, *Mosses, Liverworts, and Hornworts* (Cornell University Press, 2016). Also check out Ian Atherton et al, ed., *Mosses and Liverworts of Britain and Ireland* (British Bryological Society, 2010), and Michael Lueth's amazing photographic atlas, the *Mosses of Europe* (Poppen & Ortmann, 2019).

For general works on bryology, our top ten include: Bill and Nancy Malcolm, *Mosses and Other Bryophytes, an Illustrated Glossary* (Micro-Optics Press, 2nd. ed. 2006); Howard Crum, *A Focus on Peatlands and Peat Mosses* (University of Michigan Press, 1988) and *Structural Diversity of Bryophytes* (University of Michigan Herbarium, 2001); Alain Vanderpoorten and Bernard Goffinet, *Introduction to Bryophytes* (Cambridge University Press, 2009); and A. Jonathan Shaw and Bernard Goffinet, ed., *Bryophyte Biology* (Cambridge University Press, 2nd ed., 2009).

ONLINE RESOURCES are getting better and better. There is direct digital access to the *Flora of North America* at efloras.org; to specimens and mapped records at bryophyteportal.org; and libraries of images at, among others, the websites of Michael Lueth, the Ohio Moss and Lichen Association, Southern Illinois University, and our own Northern Forest Atlas.

There are also some fine online books. Have a look at Dale Vitt, *Field Guide to the Mosses and Liverworts of Alberta Peatlands* (NAIT Boreal Research Institute), Joannes Johanssen, *Noteworthy Mosses & Liverworts of Minnesota* (Minnesota Department of Natural Resources), Janice Glime, *Bryophyte, Ecology* (Michigan Tech) and Leica Chavoutier, *Bryophytes sl., Mosses, Liverworts, and Hornworts, Illustrated Glossary.*

PHOTOGRAPHING MOSSES is surprisingly hard. They are small, translucent, and grow packed together in dark places. At the magnifications you need to see them clearly, depth of field is less than a millimeter. They are shiny but have very little color contrast. Field lighting doesn't get the dark places, studio lighting often washes out the highlights. Imagine photographing tiny pieces of wet, chopped up lettuce in a dark-green box, and you have the idea.

There are two ways to deal with these problems. One is to ignore them and take what you get. The other is to increase dynamic range and depth of field by stacking and improve color and tonal range with lights and reflectors. We chose the second, and spent three years developing the techniques we use here. In brief, they go like this.

The close-ups, both in the field and in the studio, are shot with focus stacking. The stacks consists of anywhere from 10 to 200 images, shot by varying the point of sharp focus. We shoot with full-frame Canon DSLRS and lenses on Really Right Stuff tripods in the field and a Stackshot automated focusing rail in the studio. We light with Fotodios reflectors and portable Ikan LED panels in Chimera softboxes in the field and add larger NanGuang panels in the studio. In the field our working lenses are the Canon 24-105 mm zoom for landscapes and the 180 mm macro for close-ups. In the studio we use the Canon 55 mm MP-E and 100 mm macro lenses for routine work. For near-micro, we use Mitutoyo 5× and 10× Plan Apo microscope objectives mounted on a Canon 70-200 mm zoom telephoto.

The latter combination is magical but demanding. We routinely shoot stacks of over 100 frames, advancing the camera by as little as 0.006 mm, roughly the width of a red blood cell.

The images are combined with Zerene Stacker, and then post-processed with Lightroom and Photoshop. Sometimes they are disappointing, sometimes amazing. Either way they are a lot better than what we shot before we started stacking.

The landscape photos are shot with exposure stacking, also called HDR (high dynamic range). The stack consists of three to five images at different exposures; we combine them with Photomatix Pro and then post process with Lightroom and Photoshop.

Our methods are explained here: http://northernforestatlas.org/category/articles. A fine introduction to stacking by Rik Littlefield, who created Zerene, is: http://zerenesystems.com/presentations/4CsWallaWalla20151009/4CsWallaWalla20151009.pdf. An essential general resource is https://www.photomacrography.net.

Moss NAMES are not standardized, and as a result there are over 450 names in current use for the 350 northern-forest mosses. The choice is yours. If you like large genera, you see *Mnium* everywhere. If you like small ones you see *Mnium*, *Plagiomnium*, and *Rhizomnium*. If you like *Sphagnum* species big and easy, you talk about the *subsecundum* group. If you like them small and slippery, you talk about the sibling species *subsecundum*, *platyphyllum*, *lescurii*, *innundatum*, and *contortum*. If you like consistency from book to book, you are out of luck.

Like many field biologists, we find change to be a nuisance rather than a thrill. The names we use here are mostly the common names of the last forty years. We read the *Flora of North America*, which tends to be innovative, with interest. But we don't live our lives by it.

For those who have upgraded to newer names, here is a concordance. The names in this book are in the right column; other names in current or recent use are in the left.

Arrhenopterum heterostichum	Aulacomnium heterostichum
Atrichum altecristatum	Atrichum undulatum group
Atrichum crispulum	Atrichum undulatum group
Barbula fallax	Didymodon fallax
Barbula reflexa	Didymodon ferrugineus
Brachytheciastrum velutinum	Brachythecium velutinum
Brachythecium oxycladon	Brachythecium laetum
Brachythecium salebrosum	Brachythecium campestre
Bucklandiella venusta	Racomitrium venustum
Calliergon trifarium	Pseudocalliergon trifarium
Calliergonella lindbergii	Hypnum lindbergii
Campyliadelphus chrysophyllus	Campylium chrysophyllum
Campylophyllum hispidulum	Campylium hispidulum
Codriophorus acicularis	Racomitrium aciculare
Codriophorus fasciculare	Racomitrium fasciculare
Cratoneuron commutatum	Palustriella falcata
Dicranum scoparium in part	Dicranum bonjeanii
Drepanocladus exannulatus	Warnstorfia exannulata
Drepanocladus fluitans	Warnstorfia fluitans
Drepanocladus revolvens	Limprichtia revolvens
Drepanocladus uncinatus	Sanionia uncinata
Drepanocladus vernicosus	Hamatocaulis vernicosus
Elodium blandowii	Helodium blandowii
Elodium paludosum	Helodium paludosum
Encalypta rhabdocarpa	Encalypta rhaptocarpa
Eurhynchiastrum pulchellum	Eurhynchium pulchellum
Eurhynchium riparioides	Torrentaria riparioides
Gemmabryum caespiticium	Bryum caespiticium
Gymnostomum recurvirostrum	Hymenostylium recurvirostrum
Haplohymenium triste	Anomodon tristis
Hygroamblystegium varium	Amblystegium varium
Hygroamblystegium varium	Hygroamblystegium tenax
Hygrohypnum molle	Hygrohypnum duriusculum
Hylocomiastrum umbratum	Hylocomium umbratum
Isopterygium distichaceum	Pseudotaxiphyllum distichaceum
Isopterygium elegans	Pseudotaxiphyllum elegans
Leucodon brachypus in part	Leucodon andrewsianus
Loeskeobryum brevirostre	Hylocomium brevirostre
Mnium ambiguum	Mnium lycopodioides
Mnium appalachianum	Rhizomnium appalachianum
Mnium ciliare	Plagiomnium ciliare
Mnium cuspidatum	Plagiomnium cuspidatum
Mnium ellipticum	Plagiomnium ellipticum
Mnium hymenophylloides	Cyrtomnium hymenophylloides
Mnium medium	Plagiomnium medium
Mnium punctatum	Rhizomnium punctatum
Mnium rostratum	Plagiomnium rostratum
Niphotrichum canescens	Racomitrium canescens
Orthodicranum flagellare	Dicranum flagellare
Orthodicranum fulvum	Dicranum fulvum
Orthodicranum montanum	Dicranum montanum
Orthodicranum viride	Dicranum viride
Orthotrichum speciosum	Orthotrichum elegans
Oxyrrynchium hians	Eurhynchium hians
Oxystegus tenuirostris	Trichostomum tenuirostre
Pelekium minutulum	Cyrto-hypnum minutulum
Pelekium pygmaeum	Cyrto-hypnum pygmaeum
Rhytidiadelphus subpinnatus	Rhytidiadelphus squarrosus
Platyhypnidium riparioides	Torrentaria riparioides
Pogonatum alpinum	Polytrichastrum alpinum
Pohlia filiformis	Anomobryum julaceum
Polytrichastrum formosum	Polytrichum formosum
Polytrichastrum ohioense	Polytrichum ohioense
Polytrichastrum pallidisetum	Polytrichum pallidisetum
Polytrichum juniperinum in part	Polytrichum strictum
Polytrichastrum longisetum	Polytrichum longisetum
Ptychostomum creberrimum	Bryum lisae
Ptychostomum pseudotriquetrum	Bryum pseudotriquetrum
Pylaisiadelpha recurvans	Brotherella recurvans
Pylaisiella selwynii	Pylaisia selwynii
Racomitrium heterostichum	Racomitrium venustum group
Rosulabryum capillare	Bryum capillare
Sciaromium lescurii	Platylomella lescurii
Sciuro-hypnum plumosum	Brachythecium plumosum
Sciuro-hypnum populeum	Brachythecium populeum
Sciuro-hypnum reflexum	Brachythecium reflexum
Scorpidium revolvens	Limprichtia revolvens
Serpoleskea confervoides	Platydictya confervoides
Serpoleskea subtilis	Platydictya subtilis
Sphagnum affine	Sphagnum imbricatum
Sphagnum angustifolium	Sphagnum recurvum group
Sphagnum austinii	Sphagnum imbricatum
Sphagnum contortum	Sphagnum subsecundum group
Sphagnum fallax	Sphagnum recurvum group
Sphagnum flexuosum	Sphagnum recurvum group
Sphagnum lescurii	Sphagnum subsecundum group
Sphagnum platyphyllum	Sphagnum subsecundum group
Sphagnum subtile	Sphagnum capillifolium
Steerecleus serrulatus	Rhynchostegium serrulatum
Straminergon stramineum	Calliergon stramineum
Thuidium minutulum	Cyrto-hypnum minutulum
Thuidium pygmaeum	Cyrto-hypnum pygmaeum
Thuidium scitum	Rauiella scita
Tomenthypnum nitens	Tomentypnum nitens
Tortula ruralis	Syntrichia ruralis

INDEX

The index includes all the names of species illustrated or discussed in the text. Main entries are in bold. Synonyms are not indexed but may be found in the table on p. 165

INDEX